Methods in Molecular Biology

Series Editor
John M. Walker
School of Life and Medical Sciences
University of Hertfordshire
Hatfield, Hertfordshire, AL10 9AB, UK

For further volumes:
http://www.springer.com/series/7651

Genotyping

Methods and Protocols

Edited by

Stefan J. White

Leiden Genome Technology Center, Department of Human Genetics, Leiden University Medical Center, Leiden, The Netherlands

Stuart Cantsilieris

Department of Genome Sciences, University of Washington School of Medicine, Seattle, WA, USA

Humana Press

Editors
Stefan J. White
Leiden Genome Technology Center
Department of Human Genetics
Leiden University Medical Center
Leiden, The Netherlands

Stuart Cantsilieris
Department of Genome Sciences
University of Washington School of Medicine
Seattle, WA, USA

ISSN 1064-3745 ISSN 1940-6029 (electronic)
Methods in Molecular Biology
ISBN 978-1-4939-6440-6 ISBN 978-1-4939-6442-0 (eBook)
DOI 10.1007/978-1-4939-6442-0

Library of Congress Control Number: 2016950196

Printed on acid-free paper

This Humana Press imprint is published by Springer Nature
The registered company is Springer Science+Business Media LLC New York

Preface

The identification of sequence variation in DNA is a basic principle of genetic research. Numerous different methodologies have been developed over the past few decades, often focussed on a specific type of sequence change. The development of massively parallel sequencing approaches has made it financially and technically feasible for entire genomes to be sequenced in a rapid and cost-effective manner. Although this may seem to render many genotyping approaches obsolete, there are still a number of situations where specific, focussed assays are preferred. In this volume we have attempted to collate a broad range of different genotyping techniques.

Microsatellite analysis has many applications, including forensic identification and cell line verification. A description of a multiplex approach is provided in Chapter 1.

There may be occasions that specific sequence variants need to be genotyped. For a small number of variants in many DNA samples, High-Resolution Melt analysis (Chapter 2) and Taqman-based assays (Chapter 3) are attractive options. In situ analysis of variants in single RNA molecules is also possible (Chapter 4). For larger variant numbers, the MassARRAY system (Chapter 5) and Molecular Inversion Probes (Chapter 6) are powerful approaches.

Copy number variation (CNV) at diverse loci has been associated with a range of phenotypes, including disease. Accurate genotyping is problematic and may underlie contrasting reports in the literature. Different assays for accurately determining CNV are described here, including Pulsed Field Gel Electrophoresis (PFGE, Chapter 7), Paralogue Ratio Test (PRT, Chapter 8), Multiplex Ligation-dependent Probe Amplification (MLPA, Chapter 9), Emulsion Haplotype Fusion PCR (Chapter 10), and Droplet Digital PCR (ddPCR, Chapter 11).

In many cases a genotype alone is not sufficient information; it is also important to know on which alleles each variant is located. For combined genotyping and haplotype generation of large stretches of DNA, there are different NGS-based approaches: long range PCR combined with PacBio sequencing (Chapter 12) and Targeted Locus Amplification (TLA, Chapter 13).

Although most techniques can be applied to DNA from almost any source, some assays have been specifically optimized for certain types of organism. For bacteria, Multilocus Sequence Typing (Chapter 14) and Rapid SNP detection with pyrosequencing (Chapter 15) are described. Genotyping-by-sequencing for plant analysis is also included (Chapter 16).

Last, but certainly not least, it is critical for genotyping findings to be reported in a clear and unambiguous fashion. A summary of the most pertinent points when describing genetic variation is included (Chapter 17).

Leiden, The Netherlands
Seattle, WA, USA

Stefan J. White
Stuart Cantsilieris

Contributors

KINGSLEY K. AMOAKO • *Canadian Food Inspection Agency, National Centers for Animal Disease, Lethbridge, AB, Canada*

BARBARA ARBEITHUBER • *Institute of Biophysics, Johannes Kepler University, Linz, Austria*

JOHN A.L. ARMOUR • *School of Life Sciences, University of Nottingham Medical School, Queen's Medical Centre, Nottingham, UK*

SUZAN DE BOER • *Department of Anatomy & Developmental Biology, Monash University, Clayton, Australia*

HENK P.J. BUERMANS • *Leiden Genome Technology Center, Department of Human Genetics,, Leiden University Medical Center, Leiden, The Netherlands*

STUART CANTSILIERIS • *Department of Genome Sciences, University of Washington School of Medicine, Seattle, WA, USA*

EUNYOUNG CHAE • *Department of Molecular Biology, Max Planck Institute for Developmental Biology, Tübingen, Germany*

YUAN CHEN • *Division of Infectious Diseases, Department of Medicine, Duke University Medical Center, Durham, NC, USA*

JOHAN T. DEN DUNNEN • *Department of Human Genetics, Leiden University Medical Center, Leiden, The Netherlands; Department of Clinical Genetics, Leiden University Medical Center, Leiden, The Netherlands*

EVAN E. EICHLER • *Department of Genome Sciences, University of Washington School of Medicine, Seattle, WA, USA; Howard Hughes Medical Institute, University of Washington, Seattle, WA, USA*

JUSTINE A. ELLIS • *Murdoch Children's Research Institute, Parkville, VIC, Australia; Department of Pediatrics, University of Melbourne, Parkville, VIC, Australia*

NORIKO GOJI • *Canadian Food Inspection Agency, National Centers for Animal Disease, Lethbridge Laboratory, Lethbridge, AB, Canada*

THOMAS HAULING • *Science for Life Laboratory, Department of Biochemistry and Biophysics, Stockholm University, Solna, Sweden*

ANGELIKA HEISSL • *Institute of Biophysics, Johannes Kepler University, Linz, Austria*

EDWARD J. HOLLOX • *Department of Genetics, University of Leicester, Leicester, UK*

QUINT P. HOTTENTOT • *Department of Human Genetics, Leiden Genome Technology Center, Leiden University Medical Center, Leiden, The Netherlands*

TIMOTHY W. JANZEN • *Canadian Food Inspection Agency, National Centers for Animal Disease, Lethbridge Laboratory, Lethbridge, AB, Canada*

JACQUELINE KRÜGER • *Department of Medicine, Dermatology and Neurology, University of Leipzig, Leipzig, Germany; Department of Endocrinology and Nephrology, University of Leipzig, Leipzig, Germany; Leipzig University Medical Center, IFB Adiposity Diseases, University of Leipzig, Leipzig, Germany*

TOMASZ KRZYWKOWSKI • *Department of Biochemistry and Biophysics, Science for Life Laboratory, Stockholm University, Solna, Sweden*

RICHARD J.L.F. LEMMERS • *Department of Human Genetics, Leiden University Medical Center, Leiden, The Netherlands*

DEREK S. LUNDBERG • *Department of Molecular Biology, Max Planck Institute for Developmental Biology, Tübingen, Germany*
M. VAN MIN • *Cergentis, Utrecht, The Netherlands*
MATS NILSSON • *Department of Biochemistry and Biophysics, Science for Life Laboratory, Stockholm University, Solna, Sweden*
BENJAMIN ONG • *Murdoch Children's Research Institute, Parkville, VIC, Australia*
JOHN R. PERFECT • *Division of Infectious Diseases, Department of Medicine, Duke University Medical Center, Durham, NC, USA*
BETH A. ROWAN • *Department of Molecular Biology, Max Planck Institute for Developmental Biology, Tübingen, Germany*
DORIT SCHLEINITZ • *Department of Medicine, Dermatology and Neurology, University of Leipzig, Leipzig, Germany; Department of Endocrinology and Nephrology, University of Leipzig, Leipzig, Germany; Leipzig University Medical Center, IFB Adiposity Diseases, University of Leipzig, Leipzig, Germany*
DANELLE K. SEYMOUR • *Department of Molecular Biology, Max Planck Institute for Developmental Biology, Tübingen, Germany*
JAY SHENDURE • *Department of Genome Sciences, University of Washington School of Medicine, Seattle, WA, USA; Howard Hughes Medical Institute, University of Washington, Seattle, WA, USA*
E. SPLINTER • *Cergentis, Utrecht, The Netherlands*
HOLLY A. STESSMAN • *Department of Genome Sciences, University of Washington School of Medicine, Seattle, WA, USA*
MATTHEW C. THOMAS • *Canadian Food Inspection Agency, National Centers for Animal Disease, Lethbridge Laboratory, Lethbridge, AB, Canada*
IRENE TIEMANN-BOEGE • *Institute of Biophysics, Johannes Kepler University, Linz, Austria*
JESS TYSON • *School of Life Sciences, University of Nottingham Medical School, Queen's Medical Centre, Nottingham, UK*
ROLF H.A.M. VOSSEN • *Leiden Genome Technology Center, Department of Human Genetics, Leiden University Medical Center, Leiden, The Netherlands*
DETLEF WEIGEL • *Department of Molecular Biology, Max Planck Institute for Developmental Biology, Tübingen, Germany*
STEFAN J. WHITE • *Leiden Genome Technology Center, Department of Human Genetics, Leiden University Medical Center, Leiden, The Netherlands*

Chapter 1

Genetic Fingerprinting Using Microsatellite Markers in a Multiplex PCR Reaction: A Compilation of Methodological Approaches from Primer Design to Detection Systems

Jacqueline Krüger and Dorit Schleinitz

Abstract

Microsatellites are polymorphic DNA loci comprising repeated sequence motifs of two to five base pairs which are dispersed throughout the genome. Genotyping of microsatellites is a widely accepted tool for diagnostic and research purposes such as forensic investigations and parentage testing, but also in clinics (e.g. monitoring of bone marrow transplantation), as well as for the agriculture and food industries. The co-amplification of several short tandem repeat (STR) systems in a multiplex reaction with simultaneous detection helps to obtain more information from a DNA sample where its availability may be limited. Here, we introduce and describe this commonly used genotyping technique, providing an overview on available resources on STRs, multiplex design, and analysis.

Key words STR, Genotyping, DNA profiling, Primer design, Multiplex PCR, Capillary electrophoresis, WGA, NGS

1 Introduction

DNA fingerprinting is a genetic typing technique used to analyze the genomic relatedness between samples, to determine identity at the genetic level, and to compare DNA patterns [1]. The basis for these kinds of analyses are sequentially repeated DNA elements referred to as tandem repeats which are dispersed throughout the whole genome. There are different "classes" of tandem repeats which differ in motif size, length, and abundance (detailed reviewed by [2]). Minisatellites (also referred to as VNTRs—*v*ariable *n*umber of *t*andem *r*epeats) or more commonly microsatellites (also referred to as STRs—*s*hort *t*andem *r*epeats) are used for genetic fingerprinting (Fig. 1). VNTRs have core repeats with 9–80 bp, while STRs contain 2–5 bp repeats such as "AATG", the repeat motif in the first intron of the human tyrosine hydroxylase gene

Stefan J. White and Stuart Cantsilieris (eds.), *Genotyping: Methods and Protocols*, Methods in Molecular Biology, vol. 1492, DOI 10.1007/978-1-4939-6442-0_1, © Springer Science+Business Media New York 2017

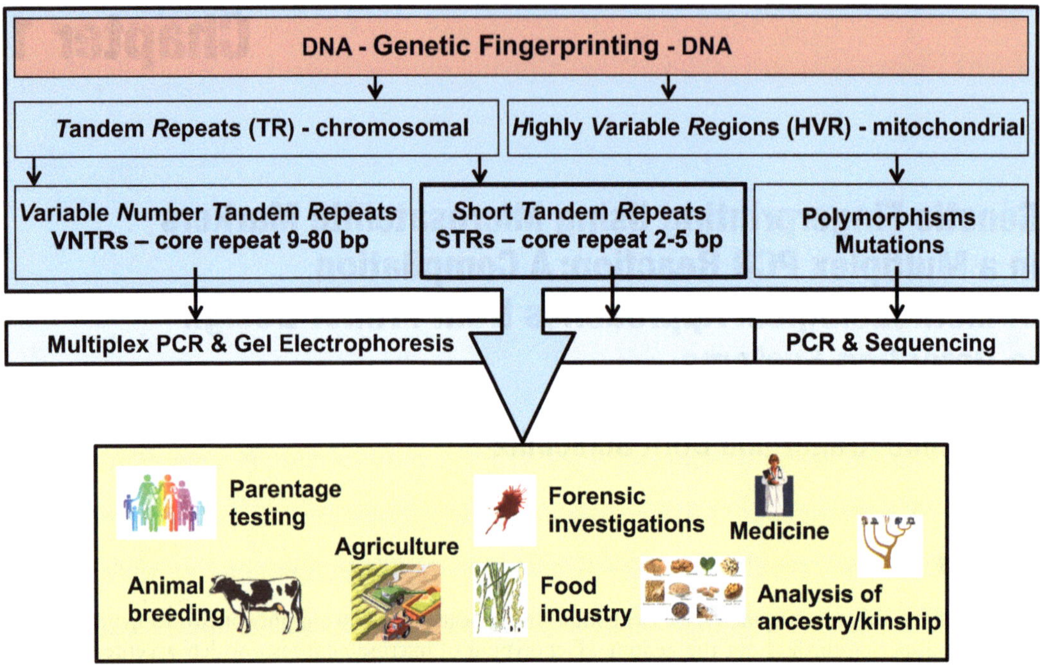

Fig. 1 Polymorphic sites in the DNA suitable for genetic fingerprinting (*dark blue*) and application areas (*yellow*)

(*THO1*) [3, 4]. The whole tandem repeat length can be several kilo base pairs in the case for VNTRs, and up to 400 bp for STRs, but numbers may vary between publications [2, 5]. The tandem repeat which is used for genetic fingerprinting needs to be polymorphic, i.e. the probability that two individuals differ in the number of repeats at their VNTR or STR needs to be high. The beginning of the human DNA-profiling era can be chronicled to 1985 when Alec Jeffreys and colleagues showed that individual-specific genetic fingerprints exist [6]. Jeffreys used hybridization probes consisting of a core sequence repeated in tandem and he visualized the samples by autoradiography. His method helped to save a young boy from deportation in 1985 [7], and to solve a criminal case in England in 1987 for the murder of two teenage girls [8]. This marked the birth of forensic DNA analysis. An individual can be homozygote for a marker, i.e. the repeat has the same length on both chromosomes or an individual can be heterozygote, i.e. there are two PCR products which differ in length. The highly characteristic DNA fingerprint pattern is not only used for forensic investigations (i.e. the identification of criminals or corpses) but also for paternity/maternity testing, analysis of kinship or in clinics to monitor bone marrow transplantation (Fig. 1) [1]. Genetic fingerprinting is also applied in agriculture (identification of plant/livestock diseases; characterization of new cultures) and in the food industry (food safety, food authenticity) [1].

It became important in the food manufacturing trade where high-value products like Basmati rice or Arabica coffee can be distinguished from inferior products [9, 10].

For human identification, even a small amount of biological material like saliva, blood, urine, sperm, hair, tissue, or bone is sufficient. The desire to gain more information from a sample, coupled with the need to limit exhaustion of a DNA sample where its availability may be limited (such as evidence obtained from a crime scene) has led to the co-amplification and typing of multiple STR systems [4]. The combination of several such systems in one PCR is referred to as multiplex reaction. Nowadays, for genetic fingerprinting in humans, usually 8–15 short tandem repeats are amplified with polymerase chain reaction (PCR) using primers flanking the region of interest. The PCR products are then subjected to flat gels (polyacrylamide, agarose) or capillary gel electrophoresis. Not only autosomal STRs can be used but also X and particularly Y-chromosomal markers have their utility e.g. in tracing paternal lineages in genealogical trees and in the forensic identification of male DNA from sexual assault cases. Alternatively, mitochondrial DNA (mtDNA) can be used if chromosomal DNA is degraded or not available in sufficient amount (low copy number (LCN) DNA<100 pg of template) (Fig. 1) [11, 12]. MtDNA is maternally inherited and the non-coding area is highly variable in sequence. A lot of effort is required to work with mtDNA as it has, except for the highly variable regions, low discriminatory power and artifacts are generated (additional fractions, contamination, loss of alleles, and loci) which can lead to inaccurate DNA profiles if an increased number of PCR cycles is used [13, 14].

The nomenclature of the STR with its corresponding alleles can be deduced from its location in genes (e.g. for *FGA*—the repeat is located in the third intron of the human alpha fibrinogen gene) or in non-coding regions (e.g. D3S1358 while D* stands for DNA and the chromosome number, and S* for the unique DNA segment specified by a number), and the number of motif repeats they contain [15, 16]. In some cases, STRs within genes and their variability have been found to be associated with serious health consequences such as Huntington's chorea [17]. The huntingtin (*HTT*) gene is located on chromosome 4p16.3 and a healthy person has 5–35 repeats of the CAG codon [18]. Expansion of this triplet results in a repetition of glutamine residues in the huntingtin protein, which accumulates gradually damaging cells in the brain. In patients with Huntington's disease, the triplet is found to be greater or equal to 36 and up to 250 times, whereby a range of 36–40 displays no full penetrance. Other such diseases include Fragile X syndrome, Friedreich's ataxia, and Machado-Joseph disease [19–21].

With the continual evolution in DNA sequencing technologies (such as NGS—next-generation sequencing), the PCR-based genotyping technique for identification in any application area is resulting in low PCR product yield. Too low T_a may possibly lead to nonspecific products caused by a high number of base pair mismatches. It is recommended to perform a temperature gradient PCR for every chosen primer pair to check for the best conditions and to evaluate if additional PCR additives like higher $MgCl_2$ concentrations or Betaine are required for optimal performance.

Third specificity:
Competition for primers exists when multiple target (organism) sequences are in a single reaction. Therefore, specificity of primer sequences should be verified which can be easily done with *B*asic *L*ocal *A*lignment *T*ools (BLAST) implemented, e.g. in the NCBI Primer-BLAST tool (http://www.ncbi.nlm.nih.gov/tools/primer-blast/) or one can use the NCBI RefSeqGene Nucleotide BLAST or the BLAST/BLAT search tool in Ensembl (http://www.ensembl.org/Multi/Tools/Blast?db=core).

Fourth primer dimer formation and "cross amplification":
The designed primers should be checked for formation of primer dimers, as this can lead to a less efficient PCR, and whether primers of different loci in the multiplex give rise to alternative amplicons. The Multiple Primer Analyzer provided by Thermo Fisher Scientific gives estimates for primer dimers and reports, e.g. T_m, CG content, extinction coefficient, and amount/OD unit (https://www.thermofisher.com/de/.../multiple-primer-analyzer.html) [42]. Another tool to test for primer dimerization is AutoDimer Software (P.Vallone©, http://www.cstl.nist.gov/biotech/strbase/AutoDimerHomepage/AutoDimerProgramHomepage.htm). Further, it is recommended to perform a primer matrix using a reference DNA template testing every primer against each other to avoid unwanted amplicons. The online tools given above are intended to give reliable guidelines as to how to create and analyze primers, but the list is not comprehensive and many other resources are available.

Fifth position of primers and composition of the multiplex:
Considering all alleles of a high polymorphic STR locus, the span between the shortest and the longest allele can add up to over 100 base pairs. Primers should flank the STR and the largest allele of one STR should not interfere with the shortest possible PCR amplicon of the following STR loci with some "space" in between. Because primers can be labeled with different fluorescent dyes which can be simultaneously analyzed, it is possible to assemble a number of STR systems in one reaction even with similar fragment length (Fig. 2). The concentration of each primer pair in the reaction needs to be balanced for an almost even peak height between the STRs so that all STR systems can be properly displayed avoiding that one dye bleeds through to another fluorescent channel. A list of ready-to-use multiplex PCR kits for genotyping is given in Table 1.

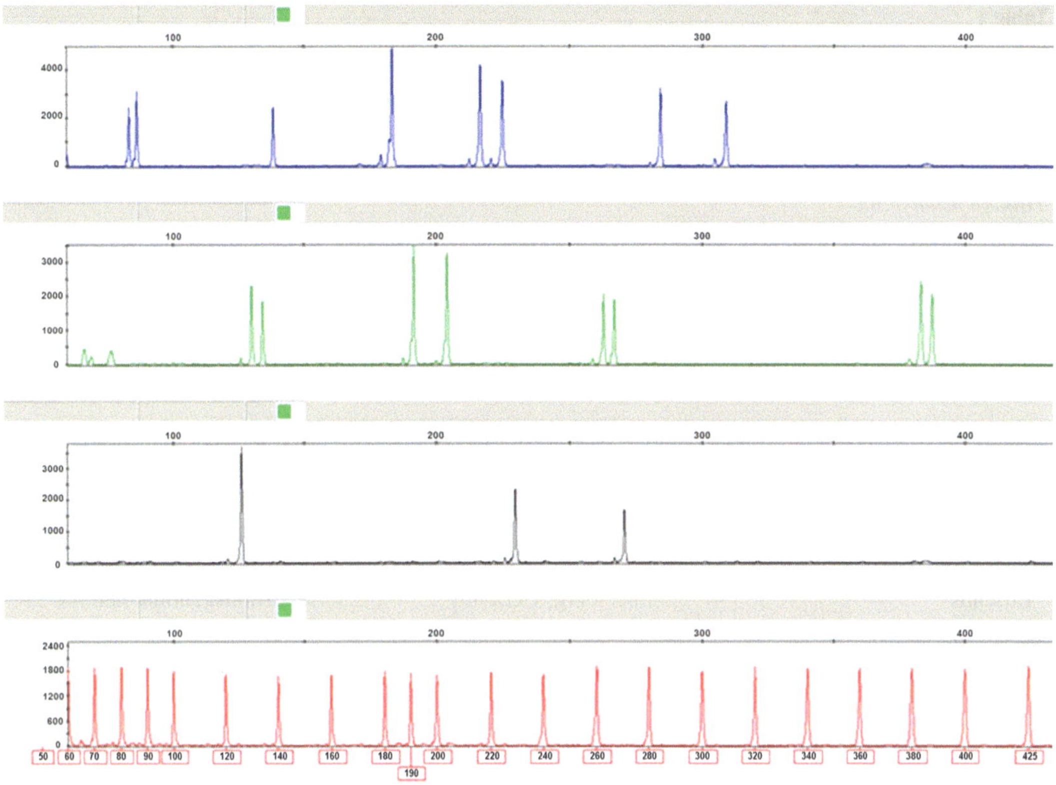

Fig. 2 Electropherogram of human STRs resolved with the ABI PRISM 310 Genetic Analyzer. Peaks in *blue*, *green*, and *black* represent the analyzed STRs in the multiplexed PCR reaction which have been labeled with different fluorophores (upper three lanes, each lane represents one fluorophore), the size standard is given in *red* (lowermost lane)

2.3 DNA Extraction and Processing for the STR Analysis

2.3.1 DNA Extraction

The procedure used in DNA isolation varies according to the type of biological sample and the nature of material where it is found [43]. An optimal procedure of DNA extraction should meet the following criteria: non-toxic, fast and cost-effective, and of course recovers highly purified DNA. There is no "universal" DNA extraction method and various commercial kits for DNA extraction have been developed providing adapted protocols for different needs [43]. All materials used for collecting biological material and DNA extraction must be sterile and free of contaminating nucleic acids. There is a remarkable report involving the murderer of a policewoman in Germany (published in daily newspapers such as "Heilbronner Phantom"). The DNA traces of a "phantom" were found at several crime scenes but more detailed analysis finally uncovered that the "trace" had its origin in contamination caused during the manufacturing process of cotton swabs. Well established kits for DNA extraction are provided by, e.g. Qiagen, Promega, or Applied Biosystems. Furthermore, particular methods are described, such as the Chelex 100 method or Phenol/Chloroform extraction [43–45].

Table 1

Selected companies providing multiplex PCR solutions, genotyping service, and optimized polymerases/master mixes for in vitro diagnostics

Multiplex PCR solutions		
Human genotyping		
Beckman Coulter	Brea, CA, USA	www.beckmancoulter.com
Biotype Diagnostic GmbH	Dresden, Germany	www.biotype.de
Ecoli PCR diagnostics.eu	Bratislava, Slowakei	www.pcrdiagnostics.eu
Promega Corporation	Madison, WI, USA	www.promega.com
Qiagen N.V.	Venlo, Netherlands	www.qiagen.com
ThermoFisher Scientific/Applied Biosystems	Foster City, CA, USA	www.thermofisher.com
Animal/agriculture genotyping		
Eurofins	Ebersberg, Germany	www.eurofinsgenomics.eu
Master Mix only		
KAPABIOSYSTEMS	Wilmington, USA	www.kapabiosystems.com

Information partly adopted from http://www.cstl.nist.gov/strbase/multiplx.htm. (10/16/2015_14:15CET)

2.3.2 qPCR Pre-amplification to Evaluate Concentration of the Sample

Real-time quantitative PCR (qPCR) is of great interest especially, but not exclusively, in forensics as this technique can rapidly detect low levels of DNA present in a (mixed) sample [46]. It derives its utility as generally genotyping kits require a certain amount of input DNA to be performed successfully whereas qPCR helps to decide which amount of DNA is optimal for the multiplex PCR in order to avoid overloading the reaction [47]. The commercial STR kits work most efficiently at the range of between 0.1 ng and 2.0 ng. High concentrations of DNA in the reaction might lead to off-scale or split peaks or a locus-to-locus imbalance. The latter can also be seen if too little DNA is used in the reaction as well as heterozygote peak imbalance or allele drop-out. Additionally, a differential quantification of male and female DNA is possible. In the context of forensic analysis, qPCR is used to study timeline gene expression of inflammation mediators at lesions in different tissues to pinpoint the time when the injuries were sustained [48, 49]. Kits are provided, e.g. by ThermoFisher Scientific/Applied Biosystems (Quantifiler Human DNA Quantification Kit) or Zymo Research (Femto Human DNA Quantification Kit).

2.3.3 Whole-Genome Amplification

Whole-genome amplification (WGA) in theory replicates the entire DNA content of a sample and can thus help to circumvent material limitations when insufficient DNA is available for planned genetic analyses [50–52]. Several methods are available for WGA: Degenerate oligonucleotide-primed PCR (DOP-PCR); primer extension pre-amplification PCR (PEP); amplification technology based on random fragmentation of genomic DNA and conversion of the resulting fragments to PCR-amplifiable library molecules flanked by universal priming sites (GenomePlex™ WGA Kit/Sigma) [53]; multiple displacement amplification (MDA, e.g. GenomiPhi™ Amplification Kit/Amersham Biosciences, illustra GenomiPhi V2 DNA Amplification Kit/GeHealthcare, REPLI-g/Qiagen); restriction and circularization-aided rolling circle amplification (RCA-RCA); and blunt-end ligation-mediated (BL-)WGA [12, 54–59]. WGA methods are applicable in medical diagnosis (e.g. cancer analysis, prenatal diagnosis) and may also be useful in forensics where different starting material (LCN or degraded DNA) usually requires different WGA approaches. However, there are conflicting data in the literature as to whether WGA introduces bias or reflects precisely the spectrum of starting DNA [12, 50, 60–62]. Technical artifacts may occur when STR markers are analyzed such as contamination, PCR failure, preferential allele amplification, the complete absence of one allele (allele drop-out, ADO) in heterozygous loci, and the nonspecific generation of extra alleles (allele drop-in).

2.4 Analytical Detection of STRs

A number of techniques are available to resolve and detect STR alleles, which are described by Butler and colleagues [63]. PCR products/DNA fragments are separated based either on size (polyacrylamide gel electrophoresis) through different staining like silver staining or fluorescent labeling (e.g. SYBR green, incorporation of a fluorescent dye on the 5′-end of a PCR primer), or by mass (MALDI-TOF mass spectrometry) [64–68]. Agarose gels could be used as well as the resolving power is sufficient to type tetranucleotide or even dinucleotide repeats [69]. However, the number of STR systems which can be amplified in one reaction is limited. These days, automated capillary gel electrophoresis is the most commonly used technique for separating fluorescence-labeled multiplexed STR PCR products (Fig. 2). Major provider/distributors for equipment are Applied Biosystems and Promega. Table 2 summarizes advantages and disadvantages of the detection systems. Prior to the determination of a sample's genotype, an allelic ladder should be analyzed for each STR on the system used for the analysis. This is important for the exact assignment of the size of the alleles, for which mobility can be different under various conditions. MALDI-TOF MS might be an exception as it has been shown that a mass precision of 0.1% relative standard deviation (RSD), which corresponds to approximately 0.1 nucleotides could be routinely observed [67].

Table 2

Technologies for STR allele resolution

Technique	Advantages	Disadvantages
Polyacrylamide gel electrophoresis (PAGE)/ silver stain	• Silver stain less expensive than fluorescence • Native gels run faster • High resolution, denaturing PA gels used for DNA sequencing → capable of single base resolution which is perfect for separating STR alleles • No expensive instrumentation required	*Native conditions* • Single base resolution not easily achieved • Heteroduplex peaks may interfere with correctly calling alleles in multiplex PCR amplifications *Denaturing conditions* • DNA fragments separate and my travel through the gel matrix at different velocities giving a double banding pattern • Not that fast • Separate lane(s) for size standard required
Automated capillary sequencers/fluorescence labeling	• Fluorescent labeling either of forward or reverse primer enables detection of only one strand • High throughput • Detection of multiple fluorophores simultaneously	• High instrument costs • Separate channel for size standard required
Microchip capillary electrophoresis	• Rapid high-throughput separation • Reduced costs • Low-volume analysis • Integration of PCR • PCR products mixed with standard allelic ladder • Development of portable device	• Fewer peaks because of short channels
MALDI-TOF mass spectrometry	• Allelic ladder not necessarily needed • High-speed analysis • High throughput	• Limited in size range of DNA fragments to be analyzed • High instrument costs
Next-generation sequencing (NGS)	• More information obtained by analyzing combination of markers that cannot be analyzed with the standard PCR method • More precise regarding the true variation of STR loci	• High instrument costs • Sophisticated bio-statistical analysis

Information partly adopted from http://www.cstl.nist.gov/strbase/tech.htm (12/10/2015_14:31 CET) and extended

Although still relatively expensive, NGS technology has advanced to the point that it can be considered a viable platform for forensic DNA analysis [70]. A properly designed assay could yield STR information in a single analysis which surpasses that of all the currently available commercial CE-based kits combined, and provide additional information on sequence variation [71].

2.5 Application Software

Automated genotyping software solutions for human identification data, e.g. the GeneMapper *ID/ID-X* software (Applied Biosystem/ThermoFisher), the GeneMarker HID software (Softgenetics), or the TrueAllele software (Cybergenetics), are well established and available for the widely used capillary gel electrophoresis systems. The analyzed DNA fragments are assigned by comparing the sizes obtained from the unknown samples with the sizes obtained for the alleles in the allelic ladder. The analysis of STRs that have been amplified with a multiplex PCR system and subsequently sequenced (MPS data) is addressed by the open source tool TSSV (Python package) which was developed by Anvar et al. for the characterization of complex allelic variants in pure and mixed genomes [72]. Several software tools are already available for identifying (forensic) STR alleles within the NGS data. Commercial available software such as Battelle ExactID (http://www.battelle.org) claims to not only identify STRs and sequence differences but also to provide information from DNA samples including hair color, eye color, ethnicity and origin, as well as to perform mitochondrial DNA analysis and to generate usable data from mixed or degraded DNA samples. In the NextGENe software (http://www.softgenetics.com), applications for targeted STR analysis and for mitochondrial amplicon analysis are implemented. Free-to-use and online available resources such as lobSTR [73] or the STR allele identification tool – Razor (STRait Razor) [70, 74] represent a good alternative to commercial tools, not least in forensic research and forensic casework. Those tools are also applicable on NGS-generated data. However, the chemistry used for library preparation and the read length of the NGS platforms used is thereby crucial for all of these analysis tools, as alleles may go undetected when the repeat region is not fully spanned [70, 72, 73].

2.6 Concluding Remarks

The analysis of STRs for genetic identity testing is in progress in every sense. First of all, analytical systems have been advanced which enables the simultaneous detection of five, six, or even eight (Applied Biosystems 3130/3130xl, 3500/3500xL Genetic Analyzers; Promega Spectrum CE System) fluorescence dyes simultaneously. Consequently, the number of STR multiplexed in one reaction increased up to 24 loci (e.g. the PowerPlex Fusion, Promega). In addition, the upcoming NGS technology not only increases the number of markers far in excess of what can be typed by CE-based methods, but also the capacity of NGS allows for sequencing of multiple samples in one analysis through the use of barcoding [71, 75, 76]. Guidelines for the nomenclature of MPS-generated STR data are going to be discussed by the International Society for Forensic Genetics (ISFG) [77]. Finally, sequencing of the whole genome may reactivate the analysis of STR variations in general as repetitive DNA sequences show more polymorphism than single nucleotide variants, and they are important in human diseases, complex traits, and evolution [72]. To date, STR multiplex PCR systems analyzed with CE are still the gold standard. As soon as the standards required for forensic investigations will be specified for NGS and permitted at court and costs will be reduced, those methods are reasonable complements or alternatives in difficult forensic cases and genetic identity testing in general.

Acknowledgements

We would like to cordially thank Peter Kovacs, head of the research group Genetics of Obesity and Diabetes, and our colleagues for their everlasting scientific and personal support. We thank Mohammed Hankir for proofreading of this manuscript.

Funding

Jacqueline Krüger is funded by a Collaborative Research Center (B03, CRC1052) granted by the German Research Foundation (DFG). Dorit Schleinitz is funded by the Boehringer Ingelheim Foundation.

References

1. Heras J, Domínguez C, Mata E et al (2015) A survey of tools for analysing DNA fingerprints. Brief Bioinform. doi:10.1093/bib/bbv016
2. Richard G, Kerrest A, Dujon B (2008) Comparative genomics and molecular dynamics of DNA repeats in eukaryotes. Microbiol Mol Biol Rev 72:686–727
3. Ruitberg CM, Reeder DJ, Butler JM (2001) STRBase: a short tandem repeat DNA database for the human identity testing community. Nucleic Acids Res 29:320–322
4. http://www.cstl.nist.gov/strbase/intro.htm
5. Vergnaud G, Denoeud F (2000) Minisateilites: mutability and genome architecture. Genome Res 10:899–907
6. Jeffreys AJ, Wilson V, Thein SL (1985) Individual-specific fingerprints of human DNA. Nature 316:76–79
7. Jeffreys AJ, Brookfield JF, Semeonoff R (1985) Positive identification of an immigration test-case using human DNA fingerprints. Nature 317:818–819
8. Roewer L (2013) DNA fingerprinting in forensics: past, present, future. Invest Genet 4:22
9. Nagaraju J, Kathirvel M, Kumar RR et al (2002) Genetic analysis of traditional and evolved Basmati and non-Basmati rice varieties by using fluorescence-based ISSR-PCR and SSR markers. Proc Natl Acad Sci U S A 99:5836–5841
10. Missio RF, Caixeta ET, Zambolim EM et al (2011) Genetic characterization of an elite coffee germplasm assessed by gSSR and EST-SSR markers. Genet Mol Res 10:2366–2381
11. Gill P, Whitaker J, Flaxman C et al (2000) An investigation of the rigor of interpretation rules for STRs derived from less than 100 pg of DNA. Forensic Sci Int 112:17–40
12. Maciejewska A, Jakubowska J, Pawłowski R (2013) Whole genome amplification of degraded and nondegraded DNA for forensic purposes. Int J Legal Med 127:309–319
13. Warner JB, Bruin EJ, Hannig H et al (2006) Use of sequence variation in three highly variable

regions of the mitochondrial DNA for the discrimination of allogeneic platelets. Transfusion 46:554–561
14. Poetsch M, Wittig H, Krause D et al (2003) The impact of mtDNA analysis between positions nt8306 and nt9021 for forensic casework. Mitochondrion 3:133–137
15. Wain HM, Bruford EA, Lovering RC et al (2002) Guidelines for human gene nomenclature. Genomics 79:464–470
16. (1992) Recommendations of the DNA Commission of the International Society for Forensic Haemogenetics relating to the use of PCR-based polymorphisms. Forensic Sci Int. **55**:1–3
17. Goellner GM, Tester D, Thibodeau S et al (1997) Different mechanisms underlie DNA instability in Huntington disease and colorectal cancer. Am J Hum Genet 60:879–890
18. Huntington's Disease Collaborative Research Group (1993) A novel gene containing a trinucleotide repeat that is expanded and unstable on Huntington's Disease chromosomes. Cell 72:971–983
19. Kremer EJ, Pritchard M, Lynch M et al (1991) Mapping of DNA instability at the fragile X to a trinucleotide repeat sequence p(CCG)n. Science 252:1711–1714
20. Campuzano V, Montermini L, Moltò MD et al (1996) Friedreich's ataxia: autosomal recessive disease caused by an intronic GAA triplet repeat expansion. Science 271:1423–1427
21. Takiyama Y, Igarashi S, Rogaeva EA et al (1995) Evidence for inter-generational instability in the CAG repeat in the MJD1 gene and for conserved haplotypes at flanking markers amongst Japanese and Caucasian subjects with Machado-Joseph disease. Hum Mol Genet 4:1137–1146
22. Berglund EC, Kiialainen A, Syvanen A (2011) Next-generation sequencing technologies and applications for human genetic history and forensics. Invest Genet 2:23
23. Borsting C, Morling N (2015) Next generation sequencing and its applications in forensic genetics. Forensic Sci Int Genet 18:78–89
24. Bandelt H, Salas A (2012) Current next generation sequencing technology may not meet forensic standards. Forensic Sci Int Genet 6:143–145
25. Becker D, Vogelsang D, Brabetz W (2007) Population data on the seven short tandem repeat loci D4S2366, D6S474, D14S608, D19S246, D20S480, D21S226 and D22S689 in a German population. Int J Legal Med 121:78–81
26. Botstein D, White RL, Skolnick M et al (1980) Construction of a genetic linkage map in man using restriction fragment length polymorphisms. Am J Hum Genet 32:314–331
27. Jones DA (1972) Blood samples—Probability of discrimination. J Forensic Sci Soc 12:355–359
28. Kruger J, Fuhrmann W, Lichte KH et al (1968) On the utilization of erythrocyte acid phosphatase polymorphism in paternity evaluation. Dtsch Z Gesamte Gerichtl Med 64:127–146
29. Guo SW, Thompson EA (1992) Performing the exact test of Hardy-Weinberg proportion for multiple alleles. Biometrics 48:361–372
30. Fung WK, Chung YK, Wong DM (2002) Power of exclusion revisited: probability of excluding relatives of the true father from paternity. Int J Legal Med 116:64–67
31. Bär W, Brinkmann B, Budowle B et al (1997) DNA recommendations. Further report of the DNA Commission of the ISFH regarding the use of short tandem repeat systems. International Society for Forensic Haemogenetics. Int J Legal Med 110:175–176
32. Caskey CT, Edwards A (1994) DNA typing with short tandem repeat polymorphisms and identification of polymorphic short tandem repeats. U.S. Patent 5: **364**,759
33. Promega Corporation (1995) Gene PrintTM STR Systems Technical Manual
34. Akane A, Shiono H, Matsubara K et al (1991) Sex identification of forensic specimens by polymerase chain reaction (PCR): two alternative methods. Forensic Sci Int 49:81–88
35. Sullivan KM, Mannucci A, Kimpton CP et al (1993) A rapid and quantitative DNA sex test: fluorescence-based PCR analysis of X-Y homologous gene amelogenin. Biotechniques 15(636–638):640–641
36. Santos FR, Pandya A, Tyler-Smith C (1998) Reliability of DNA-based sex tests. Nat Genet 18:103
37. McKeown B, Stickley J and Riordan A (2000) Gender assignment by PCR of the SRY gene: an improvement on amelogenin. Prog Foren Genet 8:433–435
38. Shewale JG, Richey SL, Sinha SK (2000) Anomalous amplification of the amelogenin locus typed by AmpFLSTR Profiler Plus amplification kit. Forensic Sci Commun **2**
39. Dieffenbach CW, Lowe TM, Dvckslcr GS (1993) General concepts for PCR primer design. PCR Methods Appl 3:S30–S37
40. Borer PN, Dengler B, Tinoco I et al (1974) Stability of ribonucleic acid double-stranded helices. J Mol Biol 86:843–853
41. Rychlik W, Spencer WJ, Rhoads RE (1990) Optimization of the annealing temperature for DNA amplification in vitro. Nucleic Acids Res 18:6409–6412

42. Multiple Primer Analyzer, ThermoFisher Scientific https://www.thermofisher.com/de/de/home/brands/thermo-scientific/molecular-biology/molecular-biology-learning-center/molecular-biology-resource-library/thermo-scientific-web-tools/multiple-primer-analyzer.html
43. Bogas V, Balsa F, Carvalho M et al (2011) Comparison of four DNA extraction methods for forensic application. Forensic Sci Int: Genet Suppl Series 3:e194–e195
44. Walsh PS, Metzger DA, Higuchi R (1991) Chelex 100 as a medium for simple extraction of DNA for PCR-based typing from forensic material. Biotechniques 10:506–513
45. Sambrook J, Fritsch E, Maniatis T (1989) Molecular cloning: a laboratory manual, vol 2, 2nd edn. Cold Spring Harbor, Cold Spring Harbor, SL
46. Kline MC, Vallone PM, Decker AE et al (2005) Testing candidate DNA quantitation standards with several real-time quantitative PCR methods. Promega meeting. Grapevine, TX
47. Reus E (2008) Anwendungen der PCR in der forensischen DNA-Analyse. Biospektrum 7:708–710
48. Liu JY (2014) Direct qPCR quantification of unprocessed forensic casework samples. Forensic Sci Int Genet 11:96–104
49. Bai R, Wan L, Shi M (2008) The time-dependent expressions of IL-1beta, COX-2, MCP-1 mRNA in skin wounds of rabbits. Forensic Sci Int 175:193–197
50. Stranska J, Jancik S, Slavkovsky R et al (2015) Whole genome amplification induced bias in the detection of KRAS-mutated cell populations during colorectal carcinoma tissue testing. Electrophoresis 36:937–940
51. Ballantyne KN, van Oorschot RAH, Mitchell RJ (2007) Comparison of two whole genome amplification methods for STR genotyping of LCN and degraded DNA samples. Forensic Sci Int 166:35–41
52. Hawkins TL, Detter JC, Richardson PM (2002) Whole genome amplification—applications and advances. Curr Opin Biotechnol 13:65–67
53. WGA Kits, Sigma-Aldrich http://www.sigmaaldrich.com/life-science/molecular-biology/automation/whole-genome-amplification.html
54. Telenius H, Carter NP, Bebb CE et al (1992) Degenerate oligonucleotide-primed PCR: general amplification of target DNA by a single degenerate primer. Genomics 13:718–725
55. Cheung VG, Nelson SF (1996) Whole genome amplification using a degenerate oligonucleotide primer allows hundreds of genotypes to be performed on less than one nanogram of genomic DNA. Proc Natl Acad Sci U S A 93:14676–14679
56. Arneson N, Hughes S, Houlston R et al. (2008) Whole-Genome Amplification by Improved Primer Extension Preamplification PCR (I-PEP-PCR). CSH Protoc 2008: pdb.prot4921.
57. Kroneis T, El-Heliebi A (2015) Whole genome amplification by isothermal multiple strand displacement using Phi29 DNA polymerase. Methods Mol Biol 1347:111–117
58. Wang G, Maher E, Brennan C et al (2004) DNA amplification method tolerant to sample degradation. Genome Res 14:2357–2366
59. Li J, Harris L, Mamon H et al (2006) Whole genome amplification of plasma-circulating DNA enables expanded screening for allelic imbalance in plasma. J Mol Diagn 8:22–30
60. Findlay I, Ray P, Quirke P et al (1995) Allelic drop-out and preferential amplification in single cells and human blastomeres: implications for preimplantation diagnosis of sex and cystic fibrosis. Hum Reprod 10:1609–1618
61. Barber AL, Foran DR (2006) The utility of whole genome amplification for typing compromised forensic samples. J Forensic Sci 51:1344–1349
62. Spits C, Le Caignec C, de Rycke M et al (2006) Whole-genome multiple displacement amplification from single cells. Nat Protoc 1:1965–1970
63. Short Tandem Repeat DNA Internet DataBase http://www.cstl.nist.gov/strbase/tech.htm
64. Bassam BJ, Caetano-Anollés G, Gresshoff PM (1991) Fast and sensitive silver staining of DNA in polyacrylamide gels. Anal Biochem 196:80–83
65. Mansfield ES, Kronick MN (1993) Alternative labeling techniques for automated fluorescence-based analysis of PCR products. Biotechniques 15:274–279
66. Monforte JA, Becker CH (1997) High-throughput DNA analysis by time-of-flight mass spectrometry. Nat Med 3:360–362
67. Butler JM, Li J, Shaler TA et al (1998) Reliable genotyping of short tandem repeat loci without an allelic ladder using time-of-flight mass spectrometry. Int J Legal Med 112:45–49
68. Robertson JM (1994) Evaluation of native and denaturing polyacrylamide gel electrophoresis for short tandem repeat analysis. In: Bär W, Fiori A, Rossi U (eds) Advances in forensic haemogenetics, vol 5. Springer, Berlin, pp 320–322
69. White HW, Kusukawa N (1997) Agarose-based system for separation of short tandem repeat loci. Biotechniques 22:976–980

70. Warshauer DH, Lin D, Hari K et al (2013) STRait Razor: a length-based forensic STR allele-calling tool for use with second generation sequencing data. Forensic Sci Int Genet 7:409–417
71. Zeng X, King J, Hermanson S, Patel J et al (2015) An evaluation of the PowerSeq™ auto system: a multiplex short tandem repeat marker kit compatible with massively parallel sequencing. Forensic Sci Int Genet 19:172–179
72. Anvar SY, van der Gaag KJ, van der Heijden JW et al (2014) TSSV: a tool for characterization of complex allelic variants in pure and mixed genomes. Bioinformatics 30:1651–1659
73. Gymrek M, Golan D, Rosset S, Erlich Y (2012) lobSTR: a short tandem repeat profiler for personal genomes. Genome Res 22:1154–1162
74. Warshauer DH, King JL, Budowle B (2015) STRait Razor v2.0: the improved STR Allele Identification Tool—Razor. Forensic Sci Int Genet 14:182–186
75. Parson W, Strobl C, Huber G et al (2013) Evaluation of next generation mtGenome sequencing using the ion torrent personal genome machine (PGM). Forensic Sci Int Genet 7:543–549
76. King JL, LaRue BL, Novroski NM et al (2014) High-quality and high-throughput massively parallel sequencing of the human mitochondrial genome using the Illumina MiSeq. Forensic Sci Int Genet 12:128–135
77. Parson W, Ballard D, Budowle B et al (2016) Massively parallel sequencing of forensic STRs: Considerations of the DNA commission of the International Society for Forensic Genetics (ISFG) on minimal nomenclature requirements. Forensic Sci Int Genet 22:54–63

Chapter 2

Genotyping DNA Variants with High-Resolution Melting Analysis

Rolf H.A.M. Vossen

Abstract

High-resolution melting analysis (HRMA) is a simple, quick, and effective method to scan and screen PCR amplicons for sequence variants. HRMA is a nondestructive closed tube assay; after PCR, DNA melting can directly be performed on the amplified samples without any purification or separation steps. For single SNP genotyping, HRMA is an attractive alternative to Sanger sequencing, restriction enzyme analysis, and hydrolysis probes.

Key words Single nucleotide polymorphism, Variant detection, Melting curve analysis, DNA

1 Introduction

DNA melting is the process where a transition from double-stranded (ds)DNA to single-stranded (ss)DNA occurs by increasing the temperature. The thermal denaturing behavior of dsDNA is dependent on base composition; it describes the manner in which dsDNA undergoes the transition to ssDNA. Not only the GC content but also the nucleotide distribution determines how dsDNA melts. The temperature at which 50 % of all dsDNA species have become single stranded is called the melting temperature (T_m). Any sequence variant can lead to a different melting behavior and T_m, which makes it possible to detect these changes by monitoring the melting process. HRMA is usually performed on amplicons in the presence of a saturating fluorescent DNA binding dye such as LC-green Plus. It is important that the dye is saturating, which means that a dye molecule occupies every binding position in the DNA. Classical dyes like SybrGreen are not suitable for HRMA, since they are not used at a saturating concentration.

HRMA is very sensitive for detecting heteroduplexes: re-annealed opposite strands of the two alleles in which there will be one or more mismatches. Heteroduplexes will usually form during PCR in DNA samples that contain heterozygous variants.

Stefan J. White and Stuart Cantsilieris (eds.), *Genotyping: Methods and Protocols*, Methods in Molecular Biology, vol. 1492, DOI 10.1007/978-1-4939-6442-0_2, © Springer Science+Business Media New York 2017

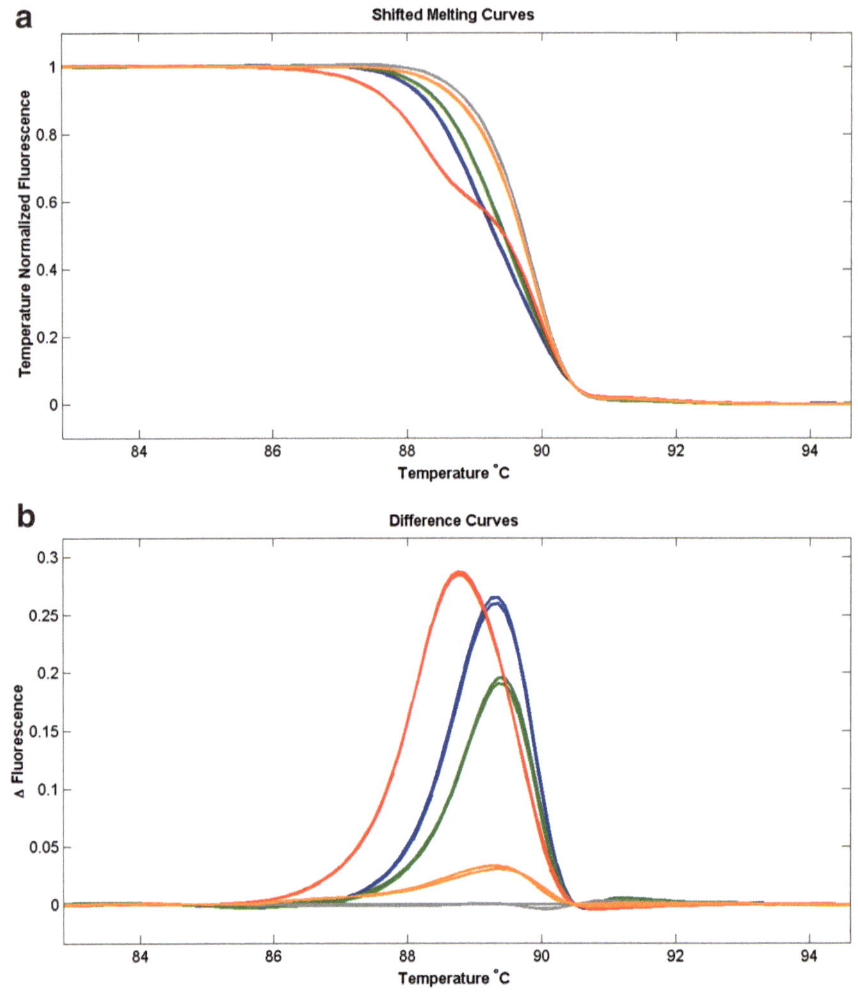

Fig. 1 (**a**) Temperature normalized melting curves, and (**b**) difference curves from DNA samples containing different sequence variants in the same amplicon

For heteroduplex detection, HRMA relies more on the shape of the melting transition then on the T_m [1] (Fig. 1). Heterozygous variants give rise to an altered melting curve and are detected with high sensitivity. In contrast, homozygous variant detection relies more on T_m change and is therefore detected with significantly lower confidence. HRMA sensitivity has also shown to be less sensitive for detecting small insertions and deletions, and these can occasionally be missed. The accuracy of HRMA depends on the instrument, the software, and the fluorescent dye being used [2].

1.1 HRMA: Scanning and Screening

Detection of unknown sequence variants in amplicons is called scanning. When a rare variant is expected in a large number of samples, scanning with HRMA will significantly reduce the workload when compared to Sanger sequencing. Known mutations can be more efficiently targeted, either by small amplicon melting or unlabeled probe melting. The workflow for scanning and screening is straightforward: after careful optimization of the PCR (*see* Subheading 1.5), samples are amplified in 96- or 384-well plates, either in real-time fashion or on a regular block thermocycler. After PCR, samples are melted and analyzed with appropriate software, which groups melting curves with overlapping profiles. HRMA is a comparative analysis: melting curves of unknowns are compared to those of a control sample which defines the baseline.

1.2 Assay Design

Scanning for unknown coding variants usually involves designing primers that cover the exons of a gene and some flanking intron sequences. Idaho Technologies has developed LightScanner primer design software that is convenient to use. It automatically generates multiple overlapping primer sets from exonic regions, thereby designing amplicons with sizes that are suitable for HRMA. For scanning and screening, amplicon size is important. In general, variants in small amplicons will be detected with greater sensitivity than in larger amplicons. The sensitivity for heterozygous variants in amplicons up to 300 bp is nearly 100% [3] and this is the maximum size recommended for diagnostic purposes. Heterozygous variants in amplicons larger than 400 bp and up to 1000 bp can be detected with a sensitivity higher than 95% [3]. When amplicon size is not important, it is best to keep the size as small as possible.

For the screening of known SNPs, it is efficient to narrow down the region of interest. This can either be done with an unlabeled probe assay or small amplicon PCR.

1.2.1 Unlabeled Probes

Unlabeled probes are convenient to use for SNP typing and the detection of small deletions [4]. In addition to the PCR primers that span a small region of 100–300 bp, a non-fluorescent 3′-blocked oligo of 20–30 nt covering the variant of interest is introduced. The location of the SNP in the probe sequence may vary but will preferably be in the middle. Better differentiation is achieved when the mismatch is in the central portion of the probe [2]. The oligo can be 3′-blocked by either a phosphate group, a dideoxy nucleotide, or an amino C3 or C6 linker. The phosphate group tends to be a less stable modification [5]. It is possible to target multiple SNPs that are in close proximity of each other with a single probe [6]. PCR with unlabeled probes is done in an asymmetric fashion: if the probe was designed on the forward strand, a 1:5–10 ratio of forward to reverse primer is used. This enhances the probe signal. One should first determine the best ratio of forward to reverse primer by testing ratios of 1:5, 1:10, and 1:15. Small amplicons are generally preferred, but in our hands amplicons with sizes of 100–200 bp gave better results with a probe than amplicons smaller than 100 bp.

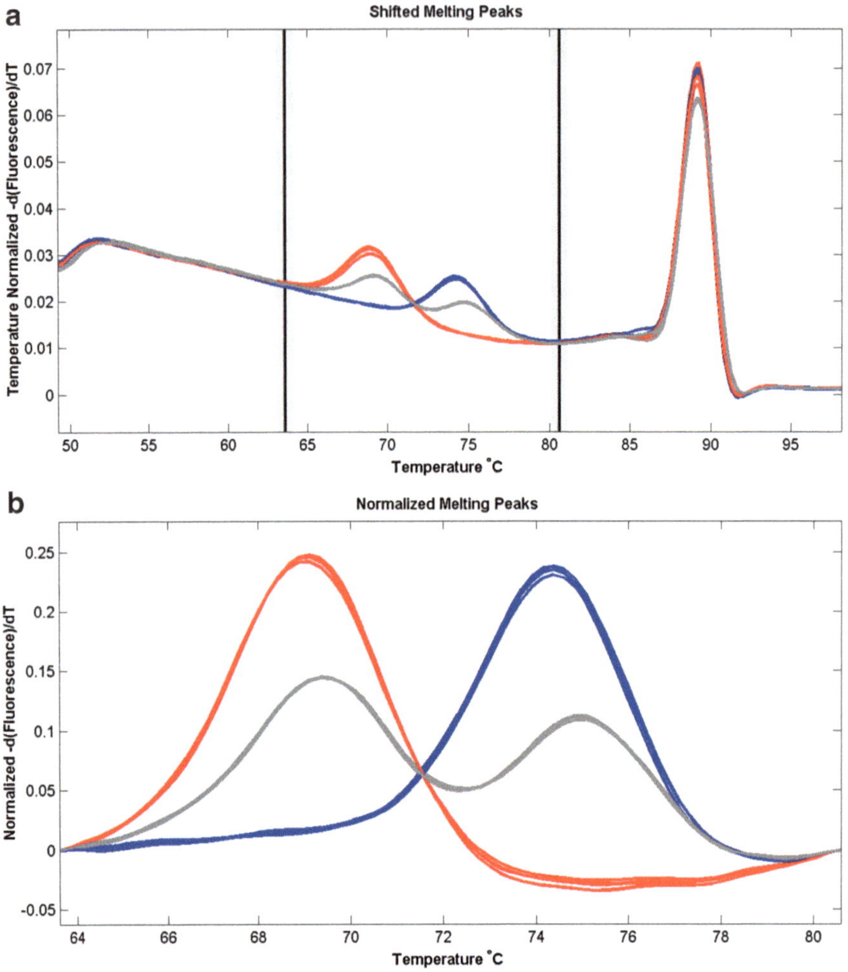

Fig. 2 (**a**) Shifted and (**b**) normalized melting peaks of an unlabeled probe assay targeting a C>G substitution. The probe sequence included the C-variant: homozygous C and thus 100 % match will have the highest T_m (*blue*). *Red* and *gray* are G/G and C/G respectively. In (**a**), both the probe melting curves in the middle and the whole amplicon melting peak at the right are seen

1.2.2 Small Amplicon PCR

For single SNP detection, one can design PCR primers directly before and after the SNP and amplify a fragment ≤50 bp [7]. All heterozygous variants will be easily detected, but resolving homozygous variants can be challenging. Especially when the GC content stays the same (e.g. G/C or A/T variants), the T_m differences will be very small and hard to detect. The use of so-called calibrator oligos is a way to enhance the resolution by minimizing the technical variability between samples [8] (Fig. 3).

1.3 Instruments for HRMA

Most real-time PCR cyclers now have the option to run an extended melting program to acquire more data points. It is the accuracy of temperature control and fluorescence measurement that defines the resolution of an instrument. The ability to measure at a data

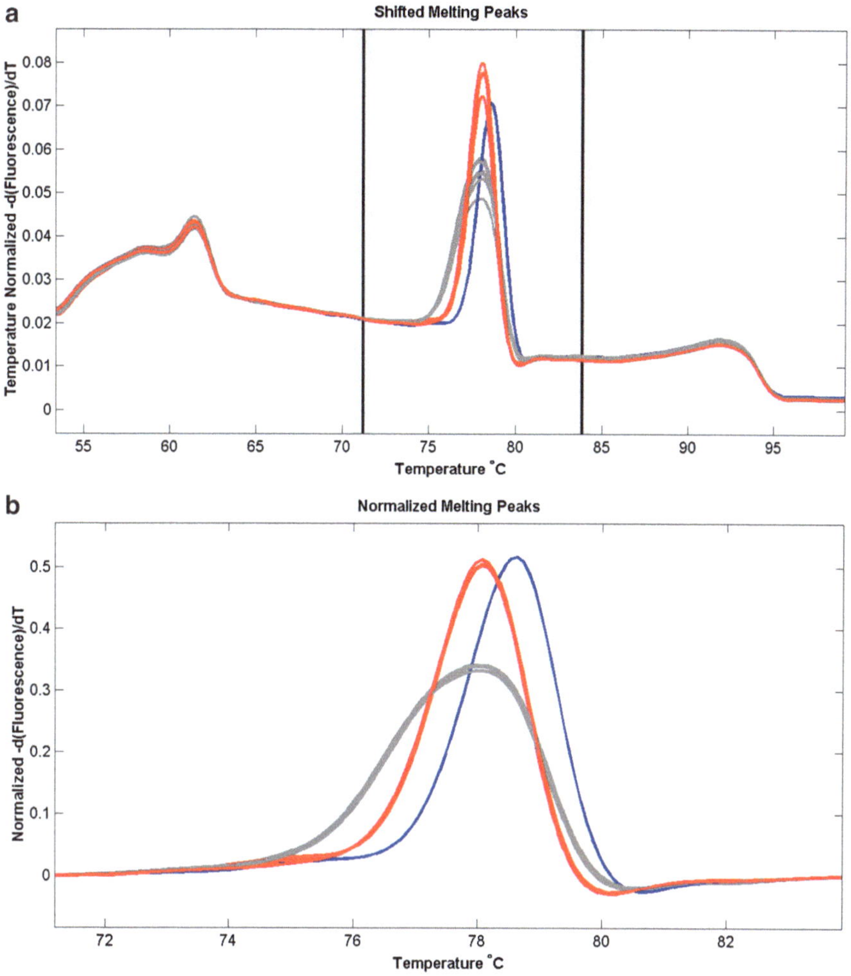

Fig. 3 Small amplicon assay, targeting a SNP and discriminating all three variants. (**a**) Shows the whole temperature range including the melting peaks from low- and high calibrator oligos from low- and high calibrator oligos (at 61 and 93 degrees respectively). (**b**) Shows the three variants. Note that the heterozygous sample (*gray*) is not resolved into two peaks. This is very common with small amplicon assays as the resolution is lower compared to unlabeled probes

density of more than 10 points/°C enhances the resolution, and is needed for HRMA [9]. Dedicated instruments for HRMA still have an advantage over general equipment [9], also because the software for those instruments is usually dedicated to HRMA and offer more analysis options. We have used the LightScanner®-96 from BioFire (formerly Idaho Technologies).

1.4 Fluorescent Dyes for HRMA

There are a few saturating DNA binding dyes available that are suitable for HRMA. We have successfully used LCGreen Plus+ (BioFire), Syto-9 (Invitrogen), and LightCycler® 480 ResoLight Dye (Roche Life Science), with a slight preference for LCGreen Plus. Others have shown Syto-9 to be comparable with LCGreen Plus [10].

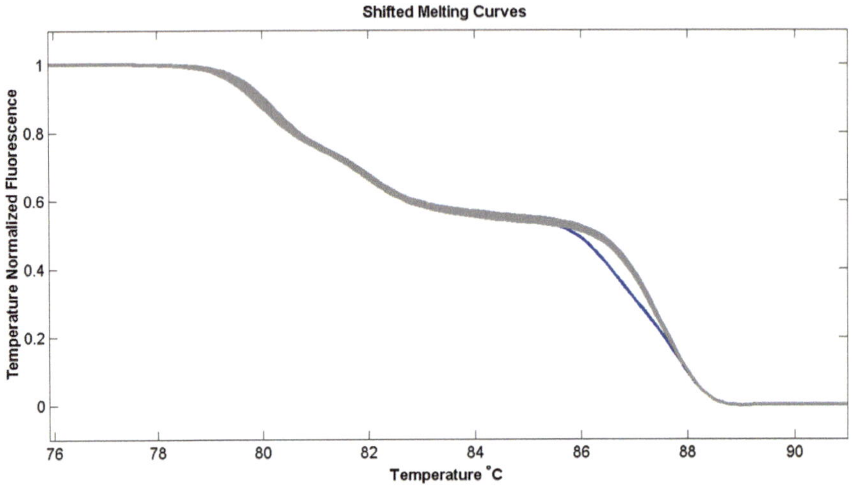

Fig. 4 Melting curve showing multiple melting domains. A variant (*blue*) is detected in the last domain

1.5 Assay Optimization

The key to success in HRMA is a well optimized PCR reaction. Any new primer design should carefully be tested with different annealing temperatures, by running a temperature gradient from, e.g. 56 °C to 68 °C. The presence of a dye such as LCGreen stabilizes DNA duplexes, and slightly raises the optimal annealing temperature. Most targets will work well at an annealing temperature of 60 °C. A well optimized amplicon gives a clean single melting peak in HRMA or band on an agarose gel. The presence of double melting domains makes it more difficult to judge the PCR conditions, as the melting curve can have more than one transitions (Fig. 4). If in any doubt, it is always useful to inspect the PCR product on a 2 % agarose gel.

Additives such as 10 % DMSO or 0.5 M Betaine can greatly improve the PCR conditions of amplicons with high GC%. Complete melting of a fragment may not be achieved due to high GC content. Addition of DMSO is then needed to lower the T_m. The melting behavior in a reaction is also dependent on the reaction chemistry and salt concentration. Different PCR mixes may give different results, and it is therefore important not to mix different chemistries in a single experiment. The salt concentration of the DNA sample also has an effect on the T_m. It is not recommended to compare DNA samples that were processed with different isolation methods, since differences in salt concentration will lead to variable results. It is also important to keep a similar amount of input DNA in all reactions, as big differences in DNA quantity will give less reproducible results.

The addition of a concentrated Tris/KCl solution can improve results that initially are variable [11] (Fig. 5). 1 μl of a Tris/KCl solution (1 M KCl, 0.5 M Tris–HCl pH 8) is added to the reactions post-PCR, followed by incubation of 2 min at 95 °C. After

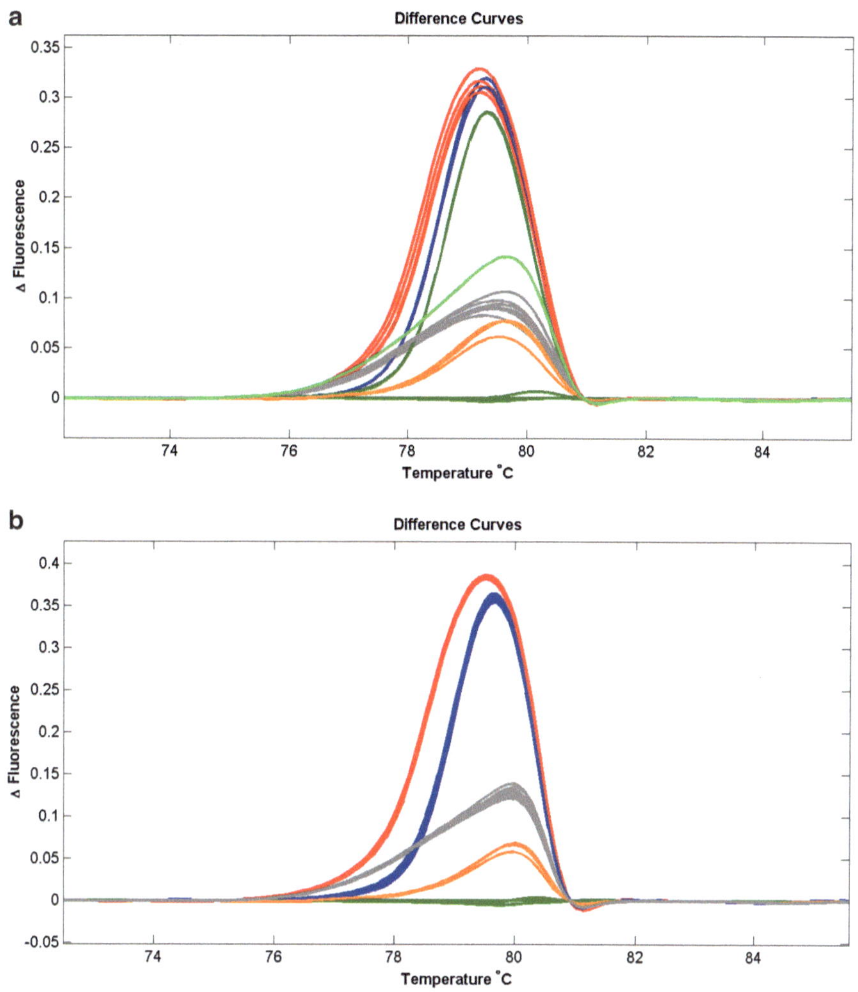

Fig. 5 Effect of the addition of a Tris/KCl solution. (**a**) Before addition of Tris/KCl. (**b**) After addition of Tris/KCl. Identical melting curves cluster much better after the addition of the solution

cooling, melting is repeated. Unfortunately, the effect can be slightly unpredictable: in some assays it will work while in other cases no improvement is seen.

2 Materials

Although several sources recommend the use of HPLC purified PCR primers, we have obtained excellent results with standard desalted oligos.

2.1 Consumables for Use with the LightScanner®-96

1. FrameStar™ 96-well skirted plates (black frame/white well, 4titude).
2. Aluminum or plastic foils.

3. Mineral oil, PCR reagent (Sigma-Aldrich).
4. FastStart Taq Polymerase (5 U/μl, Sigma) with 10× PCR reaction buffer and 20 mM $MgCl_2$ (*see* **Note 1**).
5. LCGreen Plus+ (BioFire) (*see* **Note 2**).
6. Optional: Calibrator oligos.
   Low calibrator oligo: TTAAATTATAAAATATTTATAATAT TAATTATATATATATAAATATAATA-Amine-C6
   High calibrator oligo: GCGCGGCCGGCACTGACCCGA GACTCTGAGCGGCTGCTGGAGGTGCGGAAGCGGAGG GGCGGG-Amine-C6
7. Optional: Tris/KCl solution: 1 M KCl, 0.5 M Tris–HCl pH 8.

2.2 Equipment

1. HRMA instrument, e.g. LightScanner®-96 (BioFire).
2. Thermocycler.
3. Centrifuge for spinning 96-well plates.

3 Methods

Ideally, DNA samples should be diluted to the same concentration, e.g. 10 ng/μl (*see* **Note 3**).

3.1 PCR for Scanning and Small Amplicon Analysis

1. Set up the PCR for scanning or small amplicon analysis, preparing the following mix for one reaction (10 μl reaction volume):
   - 1 μl 10× PCR-buffer 20 mM $MgCl_2$
   - 0.2 μl dNTPs (10 mM)
   - 0.3 μl F-primer (10 pmol/μl)
   - 0.3 μl R-primer (10 pmol/μl)
   - 1 μl LC-Green Plus (*see* **Note 4**)
   - 0.1 μl FastStart-Taq DNA Polymerase
   - add H_2O to 8 μl

   Optional: 0.1 μl low and/or high calibrator oligos (10 pmol/μl), for small amplicon analysis only.
2. Pipet 15 μl mineral oil in the wells of a white 96-well plate, and add 8 μl PCR-mix below the oil. Add 2 μl DNA (10 ng/μl) and seal the plate with an aluminum or plastic foil. Spin the plate briefly in a plate centrifuge.
3. Perform the following PCR program:

   10 min 95 °C

   40 cycles: 20 s 95 °C

   30 s 60 °C (*see* **Note 5**)

   40 s 72 °C

5 min 72 °C

1 min 95 °C (final denaturation before cooling to RT, stimulates heteroduplex-formation)

Cool to room temperature

3.2 PCR for Unlabeled Probe Analysis

1. Setup the PCR reaction (mix for one reaction, 10 μl reaction volume) with a 1:5 forward to reverse primer ratio.
   - 1 μl 10× PCR-buffer 20 mM MgCl2
   - 0.2 μl dNTPs (10 mM)
   - 0.1 μl F-primer (10 pmol/μl)
   - 0.5 μl R-primer (10 pmol/μl)
   - 0.5 μl probe (10 pmol/μl)
   - 1 μl LC-Green Plus
   - 0.1 μl FastStart-Taq DNA Polymerase
   - add H2O to 8 μl
2. Pipet 15 μl mineral oil in the wells of a white 96-well plate and add 8 μl PCR-mix below the oil. Add 2 μl DNA (10 ng/μl) and seal the plate with an aluminum or plastic foil. Spin the plate briefly in a plate centrifuge.
3. Perform the following PCR program:

   10 min 95 °C

   55 cycles: 20 s 95 °C

   30 s annealing temperature

   40 s 72 °C

   5 min 72 °C

   1 min 95 °C

   Cool to room temperature

3.3 Melting Acquisition and Data Analysis

After PCR, melting is performed in a machine capable of doing HRMA. In the LightScanner®-96, melting is performed at a rate of 0.1 °C/s. The temperature range at which melting is performed can vary per target and assay type. For scanning and small amplicon analysis, one can start with a broad temperature range of 60 °C–98 °C, which enables complete melting acquisition for most targets. Once the melting transition for a certain target is known, one can set a more precise temperature range to shorten the time that is needed for data collection. Unlabeled probes will dissociate earlier than amplicons and a lower starting temperature is needed, e.g. 55 °C. When using low calibrator oligos, the starting temperature can be as low as 50 °C. After data collection, melting curves are normalized by selecting a linear region before and after the melting transition (Fig. 6). Finally, temperature shifting of melting curves is done to eliminate temperature differences between samples [12] (Fig. 7). Data analysis of melting data can be quite intuitive and one has to experiment with the parameter settings to achieve the best grouping of identical curves.

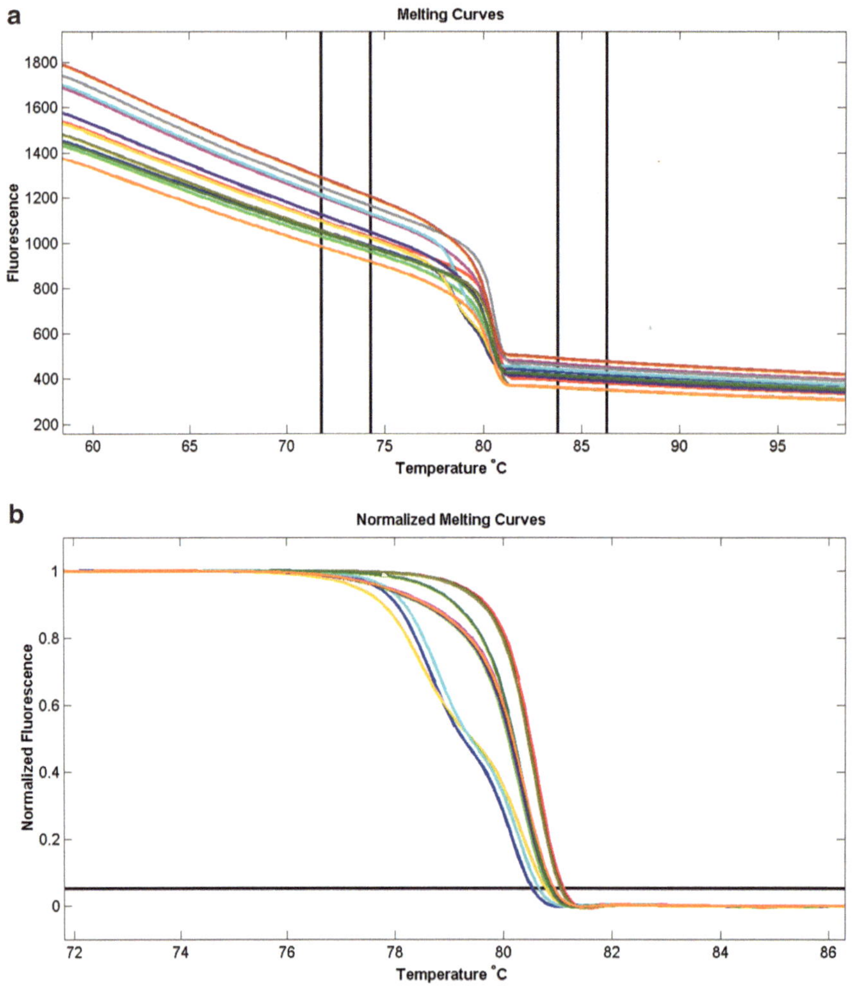

Fig. 6 (**a**) Selection of linear region before and after the melting transition. (**b**) Melting curves after normalization

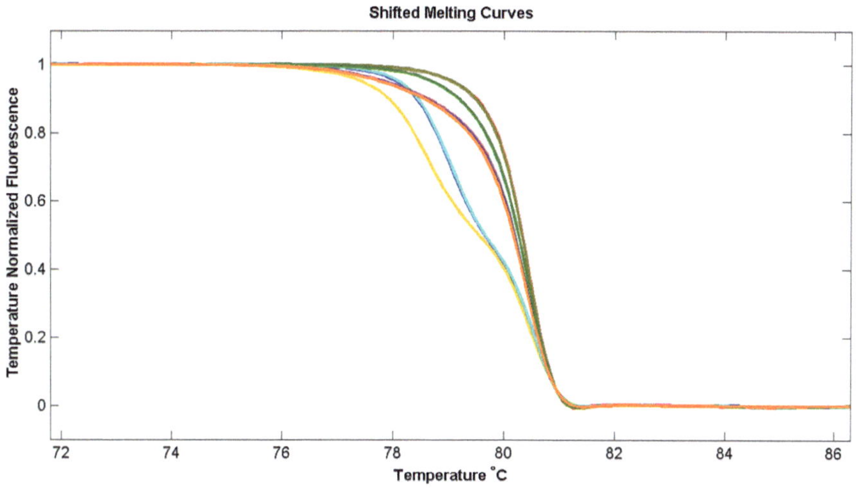

Fig. 7 Temperature shifted melting curves

4 Notes

1. The use of a hot-start Taq DNA polymerase is strongly recommended.
2. The addition of LCGreen Plus may lead to different optimal PCR conditions: re-optimization of a previously working PCR is often needed. The optimal $MgCl_2$ concentration for most targets is 2 mM.
3. When the experimental setup will allow for it, running technical duplicates is always a good idea, especially when there are differences in the amount or the quality of the DNA.
4. It is possible to add LC-Green post-PCR to an already working PCR. This is only recommended for small-scale experiments, since an extra step is added. Furthermore, adding LC-Greens will increase variation due to small differences in pipetting volumes. To add LCGreen Plus, mix 9 μl PCR product with 1 μl LCGreen Plus, incubate 3 min at 95 °C, and cool to room temperature.
5. When optimizing the PCR for many different fragments, a touch-down PCR could be considered, saving the work that is needed to optimize every fragment individually. During touch-down PCR, the annealing temperature is gradually lowered in every cycle. As an example, across the range of 40 PCR cycles one could start with 65 °C and end with 53 °C.

References

1. Zhou L, Wang L, Palais R et al (2005) High-resolution DNA melting analysis for simultaneous mutation scanning and genotyping in solution. Clin Chem 51:1770–1777
2. Erali M, Voelkerding KV, Wittwer CT (2008) High resolution melting applications for clinical laboratory medicine. Exp Mol Pathol 85:50–58
3. Reed GH, Wittwer CT (2004) Sensitivity and specificity of single-nucleotide polymorphism scanning by high-resolution melting analysis. Clin Chem 50:1748–1754
4. Zhou L, Myers AN, Vandersteen JG et al (2004) Closed-tube genotyping with unlabeled oligonucleotide probes and a saturating DNA Dye. Clin Chem 50:1328–1335
5. Cradic KW, Wells JE, Allen L et al (2004) Substitution of 3′-phosphate Cap with a carbon-based blocker reduces the possibility of fluorescence resonance energy transfer probe failure in real-time PCR. Clin Chem 50:1080–1082
6. Vossen RHAM, Duijn M, Daha MR et al (2010) High-throughput genotyping of mannose-binding lectin variants using high-resolution DNA-melting analysis. Hum Mutat 31:E186–E193
7. Liew M, Pryor R, Palais R et al (2004) Genotyping of single-nucleotide polymorphisms by high-resolution melting of small amplicons. Clin Chem 50:1156–1164
8. Gundry CN, Dobrowolski SF et al (2008) Base-pair neutral Homozygotes can be discriminated by calibrated high-resolution melting of small amplicons. Nucleic Acids Res 36:3401–3408
9. Herrmann MG, Durtschi JD, Wittwer CT, Voelkerding KV (2007) Expanded instrument comparison of amplicon DNA melting analysis for mutation scanning and genotyping. Clin Chem 53:1544–1548
10. Eijk R, Puijenbroek M, Chhatta AR et al (2010) Sensitive and specific KRAS somatic mutation analysis on whole-genome amplified DNA from archival tissues. J Mol Diagn 12:28–34
11. Vossen RHAM, Aten E, Roos A et al (2009) High-resolution melting analysis (HRMA)—more than just sequence variant screening. Hum Mutat 30:860–866
12. Herrmann MG, Durtschi JD, Bromley LK et al (2006) Amplicon DNA melting analysis for mutation scanning and genotyping: cross-platform comparison of instruments and dyes. Clin Chem 52:494–503

Chapter 3

High-Throughput Genotyping with TaqMan Allelic Discrimination and Allele-Specific Genotyping Assays

Angelika Heissl, Barbara Arbeithuber, and Irene Tiemann-Boege

Abstract

Real-time PCR-based genotyping methods, such as TaqMan allelic discrimination assays and allele-specific genotyping, are particularly useful when screening a handful of single nucleotide polymorphisms in hundreds of samples; either derived from different individuals, tissues, or pre-amplified DNA. Although real-time PCR-based methods such as TaqMan are well-established, alternative methods, like allele-specific genotyping, are powerful alternatives, especially for genotyping short tandem repeat (STR) length polymorphisms. Here, we describe all relevant aspects when developing an assay for a new SNP or STR using either TaqMan or allele-specific genotyping, respectively, such as primer and probe design, optimization of reaction conditions, the experimental procedure for typing hundreds of samples, and finally the data evaluation. Our goal is to provide a guideline for developing genotyping assays using these two approaches that render reliable and reproducible genotype calls involving minimal optimization.

Key words Real-time PCR, 5′ endonuclease assay, TaqMan assay, Dual-labeled probes, SYBR Green I method, Allelic discrimination, SNP genotyping, Allele-specific genotyping, Short tandem repeats, Microsatellites

1 Introduction

In the mid-1980s, the polymerase chain reaction (PCR) was developed by Mullis and coworkers, and to date represents one of the most powerful methods for the detection and quantification of DNA [1]. Higuchi and colleagues pioneered PCR detection by developing the first "real-time" PCR (rtPCR) system [2, 3]. In the early years of PCR, genotyping was only possible, using very laborious methods such as Southern-, dot-, or reverse dot plots [4, 5]. In 1991, the first 5′ endonuclease assay with ^{32}P-labeled probes was described by Holland et al. [6], based on the principle that a perfectly matched probe is degraded by the polymerase while a mismatched probe stays intact. The detection of the fragmented versus intact ^{32}P-labeled probes was carried out with thin layer

Stefan J. White and Stuart Cantsilieris (eds.), *Genotyping: Methods and Protocols*, Methods in Molecular Biology, vol. 1492, DOI 10.1007/978-1-4939-6442-0_3, © Springer Science+Business Media New York 2017

chromatography [6]. Holland's principle for genotyping is still being used today, but using a more accessible and high-throughput detection system based on rtPCR (known as TaqMan assays).

In rtPCR, the amplification of templates can be monitored in real-time by the incorporation of intercalating dyes or binding of labeled probes with a device that consists of a combination of a thermal cycler, an excitation light source (e.g. LED or laser), a fluorescence detection device, and appropriate software for processing the data [2, 3]. An amplification curve is obtained by plotting the increase in fluorescence during each cycle versus the cycle number, representing a very powerful detection method compared to post-PCR gel electrophoresis. A melting curve at the end of the PCR protocol helps evaluating the amplicons [7]. With the advent of rtPCR, genotyping assays have boomed indicated by the doubling of publications involving rtPCR as method for genotyping, from 2002 to 2006 [8]. Here, we discuss the use of rtPCR for genotyping polymorphisms in DNA.

The rtPCR genotyping assays are classified into two groups, based on the fluorescence moiety and the specificity of the method: (1) intercalating dyes and (2) fluorophore-labeled oligonucleotides, the second subdivided further into (a) primer probes, (b) hydrolysis and hybridization probes, and (c) nucleic acid analogs (reviewed in [9]). Given that intercalating dyes like SYBR Green I or EvaGreen bind to the minor groove of the double-stranded DNA during amplification [10], nonspecific products and primer dimers can also result in a fluorescence signal [11]. Intercalating dyes require a melting curve analysis for the identification of PCR products. By slowly increasing the temperature, the fluorescence signal decreases sharply when the melting point of a double-stranded DNA is reached. When plotting the change in fluorescence signal per temperature ($-\Delta F/\Delta T$) against the temperature, a melting peak analysis can be performed where sharp peaks represent primer dimers, nonspecific products as well as the desired PCR product [7].

In contrast, fluorophore-labeled oligonucleotides only monitor the amplification of the specific target. Basically, they can be designed in different ways (e.g. as TaqMan probes or Molecular Beacons), but the detection principle is nearly the same. Fluorophore-labeled oligonucleotides use fluorescence resonance energy transfer (FRET) [12, 13], where a donor/reporter dye gets excited and emits light with a longer wavelength. The acceptor or quencher, of which the absorption spectrum overlaps with the emission spectrum of the donor/reporter dye, absorbs the emitted light of the donor/reporter (*see* Fig. 1). The energy freed when the acceptor or quencher returns to the ground state can be converted into heat (FRET quenching) [14] in case of a quencher, or be emitted as fluorescence in case of an acceptor fluorescent dye [12, 13].

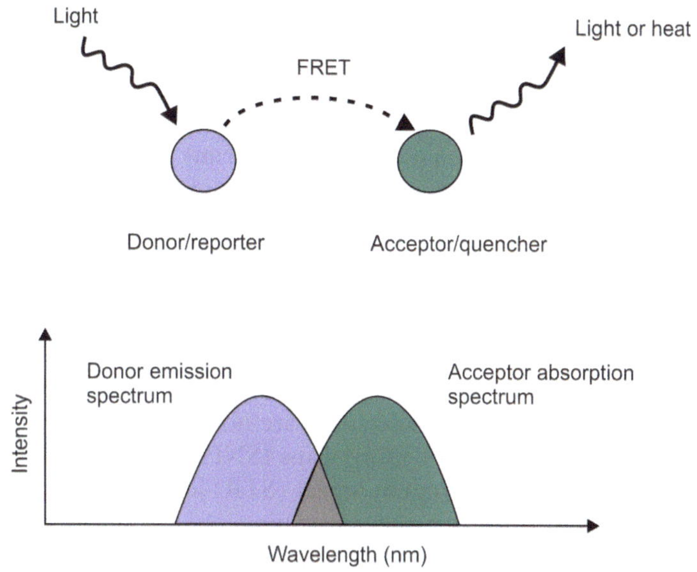

Fig. 1 Principle of the fluorescence energy transfer (FRET) between two dyes. When two fluorophores with overlapping emission and absorption spectra are in close proximity (10–100 Å), then the FRET phenomenon occurs. Specifically, a donor/acceptor dye (*purple*) gets excited and emits light with a slightly longer wavelength. Due to the overlap of both spectra (*grey shaded area*), the acceptor or quencher (*green*) absorbs the energy in form of FRET of the donor/acceptor and emits it as light, or as heat in case of a black hole quencher

The distance between the two dye moieties has to be between 10 and 100 Å (3–30 bp) for an efficient FRET [15, 16].

The features of the different fluorophore-labeled oligonucleotide subclasses are the following: the "primer-probe" subclass are fluorescently labeled primers which also act as a probe. This group includes hairpin primer-probes such as Scorpion [17] or LUX primer-probes [18], and Cyclicons [19] or Angler primer-probes [20]. Primer-probes require a melting curve analysis to monitor the formation of possible primer dimers or nonspecific products. The "hydrolysis and hybridization probe" subclass contains short oligonucleotides, specific for the PCR amplicons, carrying a donor and/or an acceptor fluorophore or quencher. The hydrolysis probes (TaqMan probes) are based on the 5′–3′ exonuclease activity of the polymerase, whereas hybridization probes (e.g. Molecular Beacons) result in a signal when binding to the PCR amplicons [21]. The "nucleic acid analog" subclass contains primer/probes with a structurally modified backbone for obtaining a certain function by incorporating them into the primer/probe sequence at specific positions. Possible modifications are locked nucleic acids (LNAs) [22], peptide nucleic acids (PNAs) [23], or phosphorothioate bonds (PTOs) [24]. They are more stable in biological fluids, and show an increased binding affinity to the target (reviewed in [9]). Additionally, LNA- and PTO-modified probes cannot be degraded by polymerases with a proofreading activity [24, 25]. LNAs play an important role in TaqMan probe design. In some cases the probe length is limited (e.g. when the SNP of interest is located near a mononucleotide run) and so the melting temperature of the probe would be too low. By incorporating LNA-modified DNA bases, the T_m is increased by ~5 °C/base [26]. So it is possible to design very short probes with high melting temperatures.

Although there is a large palette of different genotyping assays, we focused in this chapter on two rtPCR-based methods that provide fairly robust genotyping calls: TaqMan and allele-specific genotyping, both used extensively for genotyping sequence polymorphisms. These assays are not restricted to the analysis of single nucleotide polymorphisms (SNPs), but can also be used to genotype short tandem repeat (STR) length polymorphisms, especially with allele-specific genotyping.

TaqMan allelic discrimination assays, or 5′ nuclease assays [27, 28], require a specific primer pair for the amplification of a 70–150 bp long DNA fragment. In addition, it requires a polymerase with a 5′–3′ exonuclease activity (but no 3′–5′ exonuclease activity), and two different dual-labeled probes, 15–20 bp long, each labeled with a different 5′ fluorophore (e.g. FAM or HEX), and an appropriate 3′ quencher (e.g. Black hole quencher 1). The polymorphism has to be located roughly at the center of the probe. Distinguishing two alleles differing at only one position is possible, because a single base mismatch at the central position of the probe is sufficient to influence the hybridization of the short probe (*see* Fig. 2b) [6, 27, 28]. During PCR, the 5′ exonuclease activity of the polymerase releases the fluorophore from the quencher only of the perfectly matched probe that is hybridized (*see* Fig. 2a). Mismatched probes are unstable and are not digested by the 5′–3′ exonuclease activity of Taq polymerases, and do not release fluorescence (*see* Fig. 2b). The TaqMan assay can be scaled up for detecting two different polymorphisms simultaneously using one flanking primer pair and four dual-labeled probes with four different donor dyes per PCR reaction, minimizing the number of reactions, and also the amount of required sample.

Allele-specific genotyping can be used as an alternative to TaqMan assays, especially in regions with repetitive sequences, where TaqMan probes might have problems with high self-compatibility and usually fail. Allele-specific genotyping compares the amplification curves of two reactions, each with its own allele-specific primer, differing mainly at the 3′ end, which overlaps with the position of the polymorphism. The second primer is common to both alleles in the reverse orientation. The primer that perfectly matches the allele at the 3′ end will be preferentially extended compared to the other primer, rendering an earlier rtPCR amplification curve (*see* Fig. 3). The genotype can be inferred by comparing the

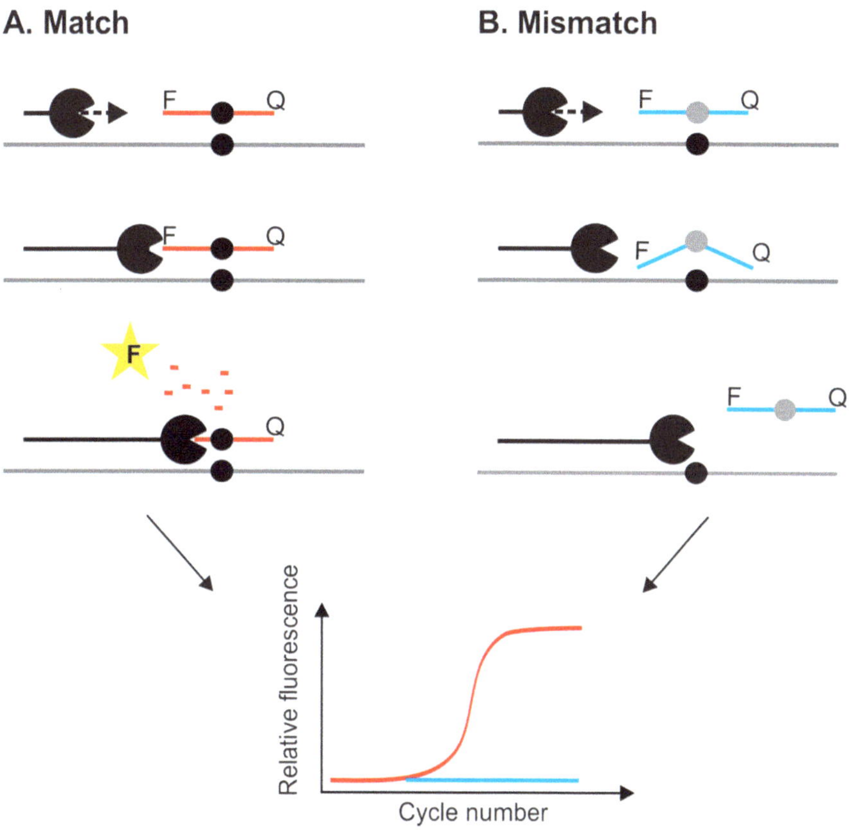

Fig. 2 Principle of a TaqMan SNP genotyping assay. (**a**) Matched probe. The polymerase (*black sphere*) elongates the primers (*black arrow-line*) and cleaves off the first 5′ base linked with the fluorophore (F) of a matching probe (in *red*), resulting in the separation of fluorophore and quencher (Q). A fluorescence signal can be measured. (**b**) Mismatched probe. In case of a mismatch, the dual-labeled probe (in *blue*) falls off before the polymerase can cleave off the first 5′ base linked with the fluorophore; no fluorescence signal is released. The PCR is carried out in a real-time thermocycler that monitors the increase of fluorescence in two different fluorescent channels at each PCR cycle

rtPCR amplification curves of the two reactions, each with its own primer pair specific for one allele. For a correct genotype calling, it is important that homozygous samples show a difference of at least 5 cycles between the amplification curves of the two primer pairs. Modifications that have an influence in the primer extension efficiency also considerably improve the accuracy of the genotype calling. For example, phosphorothioate bonds (PTOs) in the backbone of the last three bases of the 3′ end of the allele-specific primer make the primer end more rigid, such that mismatches are more inefficiently extended which enhances the difference between rtPCR curves (*see* Fig. 3) [29]. In addition, PTOs also protect the 3′ end from the exonuclease activity of the polymerases with 3′–5′ proofreading activities [24].

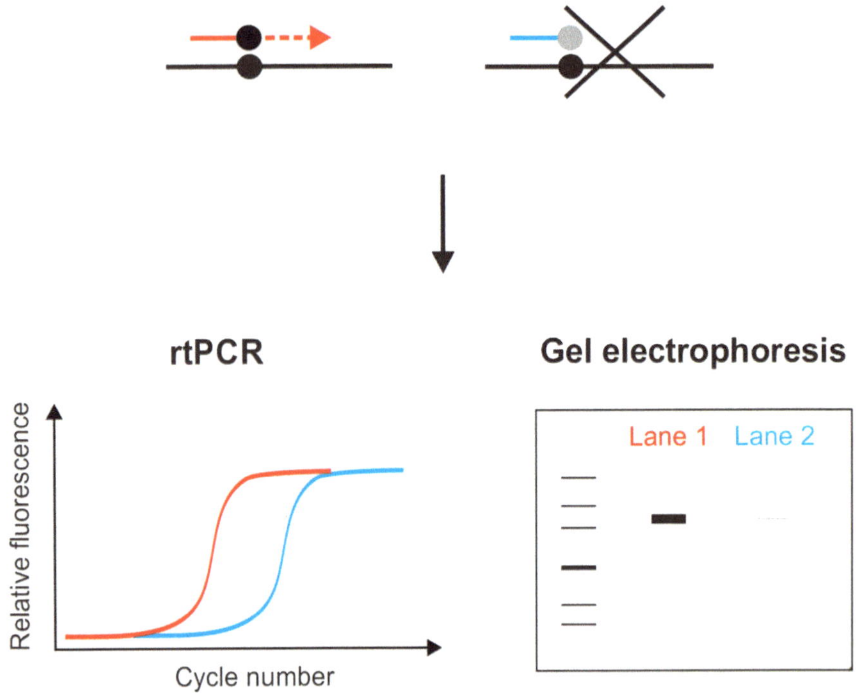

Fig. 3 Principle of the allele-specific SNP genotyping assay. Perfect matching primers at the 3′ end (*red arrow-line*) will get preferentially extended compared to a primer with a 3′ mismatch (*blue line*). The genotype can be inferred by comparing two different reactions with the same DNA template, with each containing a different allele-specific primer combination. The perfect matching primer results in an earlier amplification curve (*red curve*), or a stronger band in a polyacrylamide gel electrophoresis, when stopped before the plateau is reached, compared to the primer pair with a 3′ end mismatch in the case of a homozygous sample

The advantage of allele-specific genotyping assays is that an intercalating dye such as SYBR Green I or EvaGreen is used as the fluorophore to visualize the amplicons during PCR, which is cheaper than the TaqMan dual-labeled probes. Additionally, allele-specific genotyping is combined with melting curve analysis to verify the amplification efficiency of the product and identify possible primer dimers or nonspecific products, which is not possible with TaqMan. The major drawback of allele-specific genotyping is that for each genotype, two independent reactions are required which doubles the amount of PCR reagents, plastics and most importantly, precious sample compared to genotyping assays requiring only one reaction (like TaqMan). For both methods, optimization of the reaction conditions is required for each polymorphism, sometimes a laborious procedure that also requires gel electrophoresis to visualize possible primer dimers and nonspecific products.

The main focus of this chapter is to describe conditions that minimize the optimization procedure necessary to render reliable and reproducible genotyping calls. Since each approach has its own quirks, we structured the methods into two sections detailing in the note section important aspects specific for each method that improves the reliability of the calls and reduces the labor-intensive optimization. As an orientation for the reader, we included four different genotyping assay conditions (either for TaqMan or allele-specific genotyping) that were designed to type SNPs and STR length polymorphisms within a 3565 bp PCR product derived from a region at chromosome 16 in the human genome (chr16: 6,358,952-6,362,517, GRCh37/hg19).

2 Materials

2.1 Equipment and Plastic Consumables

1. PCR workstation with filtered airflow and UV-light (*see* **Note 1**).
2. rtPCR compatible 96-well or 384-well plates with white wells (*see* **Note 2**).
3. Optical clear rtPCR well plate seals (*see* **Note 3**).
4. rtPCR thermocycler (*see* **Note 4**).
5. Microcentrifuge with well plate rotor (*see* **Note 5**).
6. Pipettes.
7. Multichannel pipette (1–10 or 2–20 μl; 12, 24, or 64 channels).
8. Filter tips.
9. Multichannel pipetting basins (*see* **Note 6**).
10. 500 μl and 1.5 ml tubes.
11. 25 ml and 50 ml tubes for plate mastermixes.

2.2 Software

1. IDT PrimerQuest program [30].
2. Primer3Plus [31].
3. rtPCR analysis software.

2.3 TaqMan rtPCR

1. Nuclease-free water (*see* **Note 7**).
2. Hot Taq DNA Polymerase (VWR, *see* **Note 8**).
3. 10× Hot Taq Reaction Buffer S (VWR).
4. 5× Enhancer Solution P (VWR, *see* **Note 9**).
5. $MgCl_2$ (50 mM, *see* **Note 10**).
6. dNTPs (10 mM each).
7. Forward and reverse primers diluted in nuclease-free water (5 μM each; *see* **Note 11**).
8. Fluorescently dual-labeled TaqMan probes diluted in TE buffer (5 μM each, *see* **Note 11**).
9. DNA template (*see* **Note 12**).

2.4 Allele-Specific Genotyping

1. Nuclease-free water (*see* **Note 7**).
2. OneTaq Hot Start DNA Polymerase (NEB) or Phusion Hot Start II High-Fidelity DNA Polymerase (Biozym, *see* **Note 13**).
3. 5× OneTaq Hot Start Standard Buffer (NEB) or 5× Phusion Hot Start II High-Fidelity DNA Polymerase Buffer (Biozym).
4. dNTPs (10 mM each).
5. Two allele-specific primer pairs with PTO modification diluted in nuclease-free water (5 μM each; *see* **Note 14**).
6. SYBR Green I diluted in DMSO (10× stock solution) or EvaGreen fluorescent dye diluted in nuclease-free water (50x stock solution) (*see* **Note 15**).
7. DNA template (*see* **Note 12**).

3 Methods

All reactions are set up in a PCR workstation. Use nuclease-free water and filter tips (*see* **Note 1**).

3.1 Genotyping with TaqMan Allelic Discrimination rtPCR

1. In the first step, the assay needs to be designed by choosing an appropriate polymerase for your sequence and carefully designing flanking primers and dual-labeled probes. Before scaling up to genotype large amounts of samples, primer and probes have to be optimized to work appropriately. This is a critical step! The mastermix and the procedure for the optimization are the same as described below, but set up only for two samples per genotype (e.g. genotype inferred by sequencing) and non-template controls (NTC) (*see* Table 1, **Note 16**).
2. Next, set up the PCR master mix (*see* Fig. 4, Table 1, **Notes 17** and **18**). Since the 10× reaction buffer S contains only 1.5 mM $MgCl_2$ (final concentration), 1.5 mM $MgCl_2$ is added to reach a final concentration of 3 mM. Mix all components from Table 1 except the polymerase and the DNA and vortex the master mix (*see* **Note 19**). Afterward, add the polymerase and swirl the tube gently (*see* **Note 20**).
3. Aliquot 8 μl master mix into each well of the 384-well plate (*see* **Note 21**). Add 2 μl of template or water (NTC) to the aliquoted master mix using a 1–10 μl multichannel pipette (*see* **Note 22**). Cover the PCR plate with a rtPCR suitable seal (*see* **Note 23**) and spin down (*see* **Notes 24** and **25**).
4. The TaqMan rtPCR program (*see* **Note 26**) in Table 2 includes an activation step of the Hot Taq DNA Polymerase at 95 °C for 2 min followed by 45 cycles of 95 °C denaturation for 15 s and a combined annealing/extension step for 5 s depending on the melting temperature of the TaqMan probes (*see* **Note 27**).

Table 1

Mastermix for TaqMan assay with hot Taq DNA polymerase

TaqMan assay mastermix		
	1× **[μl]**	**Final conc.**
dH_2O	2.87	
10× Reaction buffer S	1	1×
5× Enhancer Solution P	2	1×
$MgCl_2$ (50 mM)	0.3	1.5 mM (total 3 mM; 1.5 mM already in the 10× reaction buffer S)
dNTPs (10 mM each)	0.2	200 μM
F+R Primer (5 μM each)	0.8	0.4 μM
FAM probe (5 μM)	0.4	0.2 μM
HEX probe (5 μM)	0.4	0.2 μM
Hot Taq DNA Polymerase (5 U/μl)	0.03	0.15 U/10 μl
Volume	8	
Aliquot 8 μl mastermix into the well plate		
DNA template	2	1500–3000 molecules
Total volume per well	10	

The 1× column denotes the volumes necessary for setting up one reaction, respectively, each with a final 10 μl volume (note that always 10% waste volume have to be considered). The usage of an Enhancer Solution P depends on the GC content of your sequence.

Add a plate read for the dyes at the correct wavelength of the dual-labeled probes (e.g. for FAM it is $\lambda_{max,\ absorption} = 494$ nm and $\lambda_{max,\ emission} = 519$ nm) to monitor the fluorescence signals and obtain an amplification curve (RFU vs. cycle number) at the end of the run (*see* **Note 28**). The final extension step is carried out for 7 min at 72 °C followed by 2 min cooling down to 25 °C.

5. Data evaluation: After the PCR run, the rtPCR software offers the possibility to show the data in an allelic discrimination plot. For the amplification plots (*see* Fig. 5a), the relative fluorescence units (RFUs), which are corrected for the background noise, are plotted on the *Y*-axis and the cycle number on the *X*-axis. By comparing the RFU values of allele 1 (*X*-axis) with the RFU values of allele 2 (*Y*-axis), the genotypes are easily distinguishable in scatter plots. The left-most cluster (blue rectangle) in Fig. 5b represents the homozygotes for allele 2 for the HEX-labeled TaqMan probe; the lower right cluster (orange

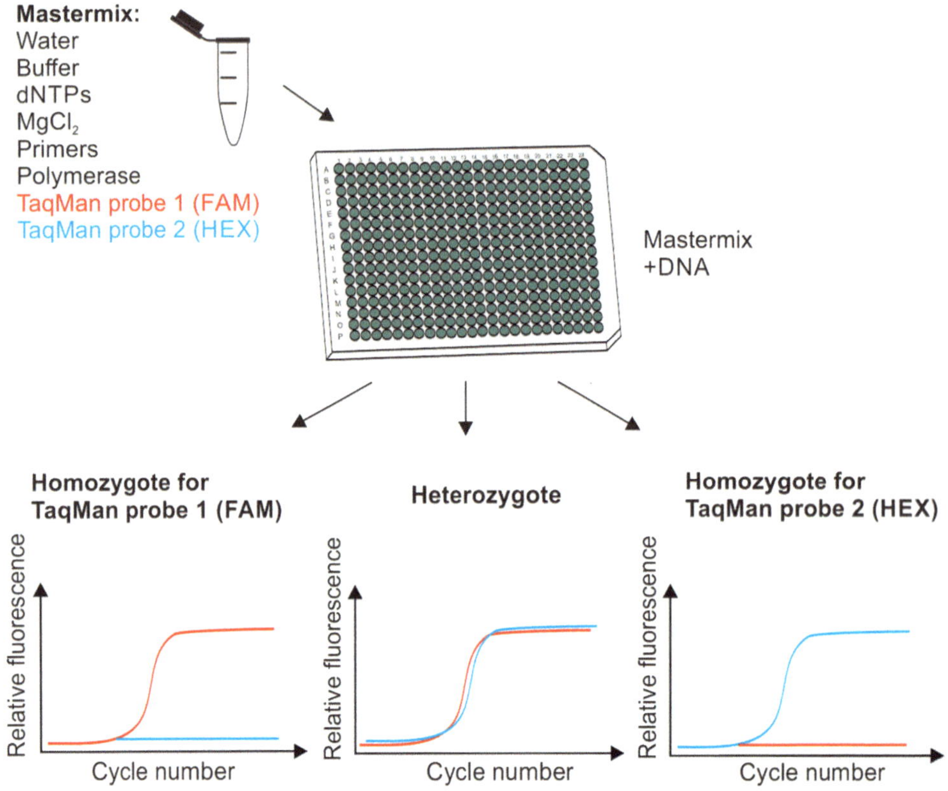

Fig. 4 TaqMan allelic discrimination assay work scheme. First, the mastermix is prepared, aliquoted into the well plate and finally the DNA samples are added. After each amplification step, the fluorescence signal specific for each allele is recorded and plotted as relative fluorescence versus cycle number (amplification plot), reflecting the genotype of the DNA sample

Table 2
rtPCR cycling protocol for TaqMan assays

TaqMan PCR cycling protocol			
Step	**Temperature**	**Time**	**Notes**
1 activation	95 °C	2 min	
2 denaturation	95 °C	15 s	Go to step 2; repeat 45×
3 annealing/extension	T_m of probes	5 s	
	Plate read for FAM and HEX		
5 final elongation	72 °C	7 min	
6 cooling down	25 °C	2 min	

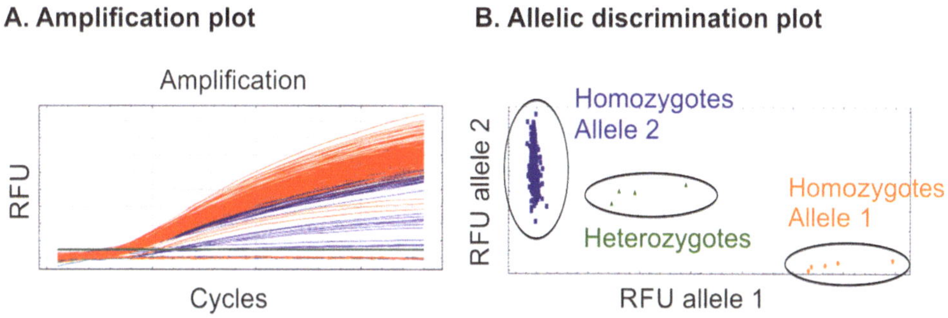

Fig. 5 TaqMan allelic discrimination data evaluation. (**a**) Amplification plot. Shows the rtPCR amplification plot for SNP rs8060928 C/T with a FAM-labeled probe for C (allele 1, *red*) and a HEX-labeled probe for T (allele 2, *blue*). (**b**) Allelic discrimination plot. This shows the allelic discrimination plot, comparing the relative fluorescence units (RFU) of both alleles. Each cluster represents a different genotype

circles) represents the homozygotes for allele 1 for the FAM-labeled probe and the cluster in the middle (green triangles) represent the heterozygotes (*see* **Notes 29** and **30**).

3.2 Genotyping with Allele-Specific rtPCR

For this assay, we used two different polymerases: OneTaq Hot Start DNA Polymerase for SNPs or Phusion Hot Start II High-Fidelity DNA Polymerase for STR or microsatellite length polymorphisms. Thus, the protocol described here is adjusted for these polymerases. However, the assay is also compatible with a wide range of other polymerases, for which reaction mixes and cycling programs need to be adapted to what is suggested by the vendor's manual (*see* **Note 13**). An overview of the different steps is shown in Fig. 6.

1. First, carefully design your allele-specific primers, choose an appropriate polymerase and an intercalating dye, and optimize them before scaling up for high-throughput genotyping. This step is critical for the whole assay! The procedure and the mastermix is the same as given below, but running only two samples for each genotype and a NTC (*see* Tables 3 or 4, **Note 31**).
2. Prepare the PCR mastermix without primers and the DNA template (*see* **Notes 17** and **32**). Mix all components from Tables 3 or 4 except for the polymerase and the DNA. Vortex the mastermix, spin the mix down and add the polymerase (*see* **Note 18**). Swirl the tube gently (*see* **Note 20**).
3. Separate the mastermix into two tubes and add to each tube one of the two different primer combinations. Mix the mastermixes gently and aliquot 5 μl into a well (*see* **Note 33**) and add 5 μl of the DNA template (*see* **Note 34**). Cover the PCR plate with a rtPCR suitable seal (*see* **Note 23**) and spin down (*see* **Notes 24** and **25**).
4. The allele-specific rtPCR program (*see* Tables 5 or 6, **Note 35**) starts with an activation step of the polymerase, followed by 45

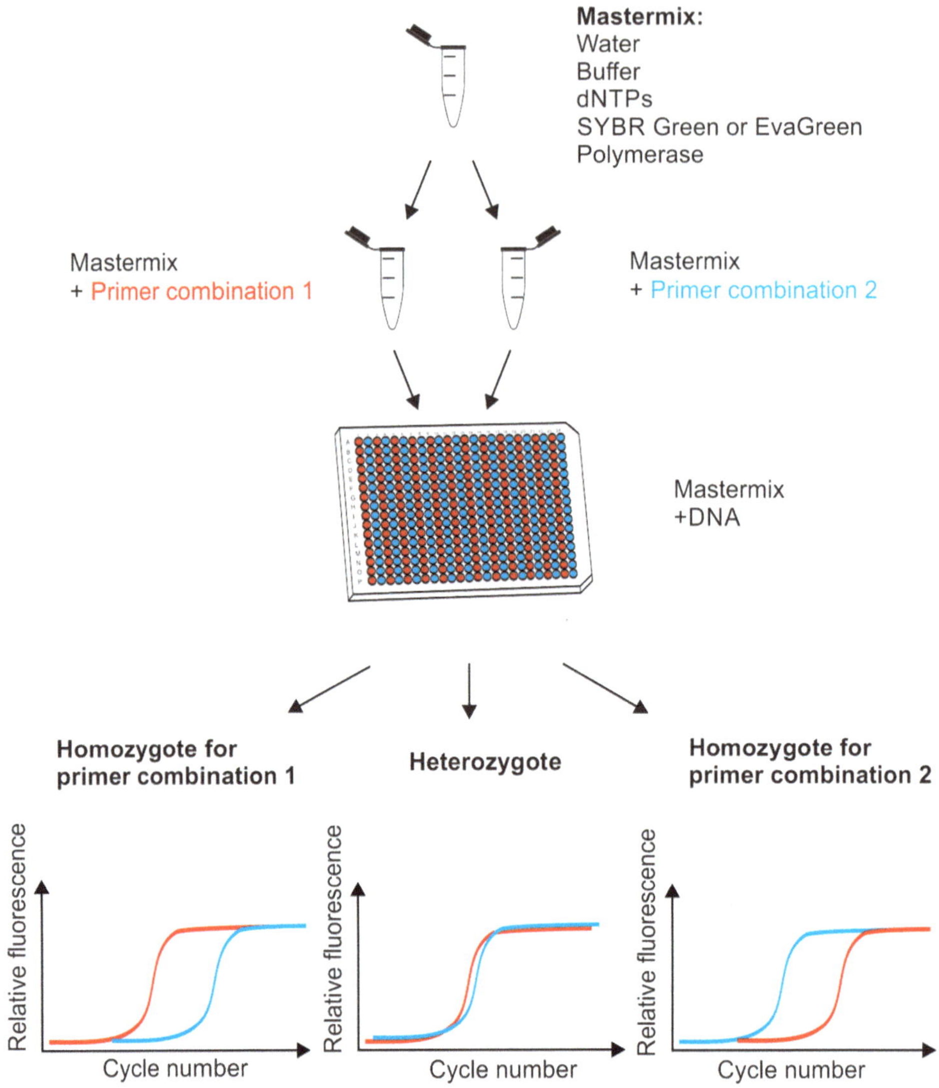

Fig. 6 Allele-specific genotyping work scheme. After the preparation of the mastermix, it is separated into two aliquots and the different primer pairs are added. Then, 5 μl of each mastermix is added in alternating order (odd and even wells) into the well plate, followed by the addition of the DNA. Note that two wells (with primer combination 1 and 2) must always include the same DNA template for comparing the amplification of both primer combinations

cycles of denaturation, primer annealing, and extension. A plate read is added to the extension step of each cycle and the wavelength of the plate read depends on the intercalating dye. In case of SYBR Green I or EvaGreen, the lighting setting with 586 nm maximum absorption and 605 nm maximum emission is used (presettings of rtPCR cycler). After a final elongation step, the PCR assay is finished with a melting curve ranging from 65 to 95 °C with a 0.5 °C increment per minute (*see* **Note 36**).

Table 3

Mastermix for allele-specific SNP genotyping with OneTaq Hot Start DNA Polymerase

Allele-specific PCR mastermix for SNP genotyping		
	1× [μl]	**Final Conc.**
dH_2O	1.875	
5× Standard Buffer	2	1×
dNTPs (10 mM each)	0.2	200 μM
SYBR Green I (10×)	0.1	0.1×
OneTaq Hot Start DNA Polymerase (5 U/μl)	0.025	0.125 U/10 μl
Volume	4.2	
Separate the mastermix into two tubes		
F+R Primer (5 μM each)	0.8	0.4 μM
Volume	5	
Aliquot 5 μl mastermix to each well		
DNA template	5	1500–3000 molecules
Total volume per well	10	

Table 4

Mastermix for allele-specific STR length polymorphism with Phusion Hot Start II High-Fidelity DNA Polymerase

Allele-specific PCR mastermix for STR length polymorphism genotyping		
	1× [μl]	**Final conc.**
dH_2O	1.85	
5× Standard buffer	2	1×
dNTPs (10 mM each)	0.2	200 μM
SYBR Green I (10×)	0.1	0.1×
Phusion Hot Start II High-Fidelity DNA Polymerase (2 U/μl)	0.05	0.1 U/10 μl
Volume	4.2	
Separate the mastermix into two tubes		
F+R Primer (5 μM each)	0.8	0.4 μM
Volume	5	
Aliquot 5 μl mastermix to each well		
DNA template	5	1500–3000 molecules
Total volume per well	10	

Table 5

rtPCR cycling protocol for allele-specific SNP genotyping assays with OneTaq Hot Start DNA Polymerase

Allele-specific PCR cycling protocol for SNP genotyping			
Step	**Temperature**	**Time**	**Notes**
1 activation	95 °C	2 min	
2 denaturation	95 °C	15 s	Go to step 2; repeat 45×
3 annealing	T_m of primer	5 s	
4 extension	68 °C	15 s	
	Plate read for SYBR Green I		
5 final elongation	72 °C	7 min	
6 melting curve	65–95 °C	0.5 °C/min	

Table 6

rtPCR cycling protocol for allele-specific short STR length polymorphism genotyping assays with Phusion Hot Start II High Fidelity Polymerase

Allele-specific PCR cycling protocol for STR length polymorphism genotyping			
Step	**Temperature**	**Time**	**Notes**
1 activation	94 °C	2 min	
2 denaturation	94 °C	15 s	Go to step 2; repeat 45×
3 annealing	T_m of primer	5 s	
4 extension	72 °C	10 s	
	Plate read for SYBR Green I		
5 final elongation	72 °C	7 min	
6 melting curve	65–95 °C	0.5 °C/min	

5. Data evaluation: The genotypes of a DNA sample can be inferred by comparing amplification of both allele-specific primer combinations. The amplification curve (allele 1: red, allele 2: blue) that is preferentially amplified (the curve which rises first, lower C_q value) represents the genotype of the template (*see* Fig. 7, **Note 37**).

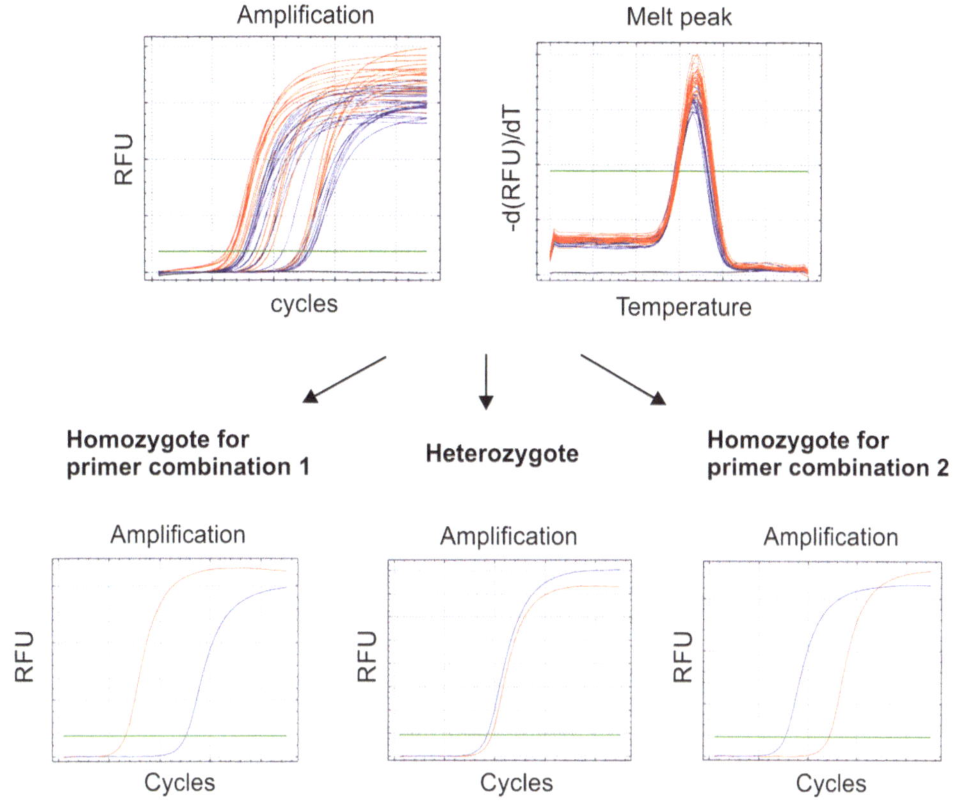

Fig. 7 Allele-specific genotyping data evaluation. This figure shows a genotyping reaction with two different primer pairs, where the *red curve* represents allele 1 and the blue curve represents allele 2. Homozygotes show a difference in amplification efficiency of about 10 cycles; for heterozygotes both amplification curves come up almost simultaneously

4 Notes

1. A fundamental procedure in performing rtPCR assays is frequent cleaning of PCR workspaces and pipettes. Therefore, we use 70 % ethanol and 10 % chlorine or DNA off (Takara) as well as UV-light. In our lab we have two separate PCR workspaces (Thermo Fisher Scientific) with filtered airflow and UV-light in separate rooms to avoid cross-contamination. In the first PCR room, we prepare stock solutions and mastermixes with polymerases, primers, dNTPs, and buffers. We also store stock solutions in this room. Additionally, in this workspace we perform genomic DNA extractions, but no PCR products are handled here. If PCR templates are genotyped, the mastermix is prepared in the first workspace and the PCR templates are added in the second PCR workspace. Never open tubes or plates containing PCR products in the same workspace used for preparing mastermixes to avoid contamination. In case

contamination does occur, clean the workspace and pipettes with 10% chlorine or DNA off cleaning solution, exchange the plastics, and turn on the UV-light. Usually 10 min of UV irradiation is sufficient. Also consider that 10% chlorine is quite corrosive (especially for shafts of pipettes) and could produce free radicals that could inhibit PCR or interfere with other experiments in the lab. These instructions are especially important if you work with single molecules.

2. We recommend PCR plates with white wells for an optimal fluorescence signal detection (FameStar or Biozym). White wells maximize the reflection of light and lead to an increase in signal-to-noise ratio. This enhances the sensitivity and reproducibility within rtPCR experiments.
3. Use optical clear rtPCR foils for fluorescence measurements. We recommend Microseal 'B' Adhesive Seals from BioRad.
4. We have good experience with the BioRad CFX 384 rtPCR cycler system. The handling is very intuitive and the CFX software offers several analysis and evaluation tools for the PCR data, but other rtPCR cyclers can also be used for genotyping assays.
5. For 96-well plates a touch spin is sufficient, but for 384-well plates a centrifugation step for 2 min at $2000 \times g$ is recommended. This is an important step to get rid of air bubbles in the reaction wells that could lead to high fluorescence background or false signals.
6. We use multichannel pipetting basins to aliquot the master mix easily into the 96- or 384-well plates.
7. Nuclease-free water is highly recommended, because nuclease contamination can lead to inconsistencies or even experiment failure.
8. There are several premixed TaqMan assays where only primer, dual-labeled probes, and DNA must be added. This has the advantage of minimizing the time for mastermix preparation and increasing the consistency between experiments, but not every polymerase performs equally for different sequences. For this purpose, we decided to use a polymerase system that is common and cheap (Hot Taq DNA Polymerase, VWR). Moreover, the polymerase should render clean PCR products with a minimal amount of nonspecific products. Using in-house mastermixes considerably reduce the costs, especially when running a high amount of genotyping reactions (approx. 500–1000 reactions per SNP). Any polymerase could be used, but the only limitations are that it must have a 5′–3′ endonuclease activity, and no 3′–5′ activity, which is part of proofreading polymerases. A proofreading activity can cause false-positive fluorescence signals if the polymerase binds directly to the probe, instead of the primer, cleaving off the 3′ base linked to the quencher and resulting in a false-positive fluorescence signal.

9. A 5× Enhancer Solution P is used for GC-rich regions. The use of this additional buffer results in slightly higher fluorescence signals within our region on chromosome 16.

10. $MgCl_2$ concentration should be between 1.5 mM and 3 mM. Higher $MgCl_2$ concentrations result in a higher efficiency in TaqMan assays. We always use 3 mM $MgCl_2$ in our assays.

11. TaqMan allelic discrimination assay design has been performed according to the BioRad Application Guidelines [32]. TaqMan flanking primer and probe design is performed with the PrimerQuest® program [30]. Primers are designed with a T_m ~50–55 °C and probes with a T_m approximately ~5–10 °C higher than that of primers [33]. TaqMan allelic discrimination probes should be rather short with about 15–20 bp. In cases where the T_m and GC content is too low, the probe is designed to be slightly longer or LNA modifications can be introduced. Probes with LNA modifications are in the range of 10–14 bp, but they are quite expensive. For comparison, probes for quantitative experiments are in a range of ~23–26 bp. The shorter the probes the better the quenching effect on the 5′ fluorescent dye, and high background noise can be avoided [33]. Careful probe design reduces probe optimization in TaqMan assays and saves costs and time to redesign new probes. Always check your primers and probes for self-complementarity, and carry out a primer BLAST to ensure that the primer is specific to your sequence and does not bind to any other site [30]. In most cases, dual-labeled probe design by online tools (e.g. PrimerQuest®) does not produce a completely satisfactory probe. For this reason we also consider the following aspects in the probe design: The 5′ base must not be a G, because G can still quench the fluorophore even after the hydrolysis of the fluorophore from the probe. Based on our experience the GC content of the probe should be between 40 and 60% and should contain more Cs than Gs. In the case that one of the two dual-labeled probes has a lower GC content depending on the type of polymorphism, the probe with the lower GC content is designed with one or two additional bases. Note, that the SNP should still be located in the middle of the probe. If you use FAM and HEX as fluorophores, we recommend to label the probe with the lower T_m/GC content or the weaker allele (A or T) with FAM, because HEX gives a slightly lower signal. This effect has been observed also by other groups [34]. Therefore also *see* **Note 29**. We recommend the use of black hole quenchers [35] instead of TAMRA. TAMRA is a fluorophore and this can result in a high background noise. Table 7 shows TaqMan genotyping primers and probes as well as PCR conditions for our region on chromosome 16.

12. In a typical genotyping reaction, we use 1500–3000 molecules (or 5–10 ng) of human genomic DNA as the template starting

Table 7

TaqMan allelic discrimination assay design

TaqMan assay design	Sequence	Primer/probe length [bp]	T_m [°C]	GC content [%]	Product length [bp]	Optimized annealing temp [°C]
Primer set 1: SNP rs8060928 C/T (genome position: 6.310.566, GRCh37/hg19)						
Forward primer	CTAACCTCTCTACCACC	17	52.9	53	130	57
Reverse primer	TGACCTCATTCAGGTGTC	18	53.7	50		
TaqMan probe 1 (**G**, antisense) 5′ FAM 3′ BHQ1	TGTCCTTGA**G**AGGACCCT	18	64.8	56		
TaqMan probe 2 (**A**, antisense) 5′ HEX 3′ BHQ1	TGTCCTTGA**A**AGGACCCT	18	63.3	50		
Primer set 2: SNP rs12102448 A/G (genome position: 6.310.773, GRCh37/hg19)						
Forward primer	GTCAAACTGTACTGTCAC	18	51.4	44	143	57
Reverse primer	CACTCTTAGAATCCAGTTAG	20	53.2	40		
TaqMan probe 1 (**G**, sense) 5′ FAM 3′ BHQ1	CAGATGTCTAC**G**AATGAAGAGT	22	56.5	41		
TaqMan probe 2 (**A**, sense) 5′ HEX 3′ BHQ1	CAGATGTCTAC**A**AATGAAGAGTC	23	57.1	39		

Primer and probes for two different regions of the human chromosome 16. The nucleotide complementary to the SNP (bold and underlined) is placed in the middle of the TaqMan probe.

number. If using PCR products as templates, the PCR products need to be diluted to reduce the starting template numbers to a maximum of 10^9–10^{10} molecules by first testing a 1:10, 1:100, and 1:1000 dilution.

13. Usually, probe design of TaqMan assays is very difficult in repetitive sequences or for SNPs which are surrounded by mono-, di-, and trinucleotides. In such cases, allele-specific genotyping is a powerful alternative. Routinely, OneTaq DNA Polymerase (NEB) is used for genotyping, but sequences with STRs (especially polyA runs) require more accurate polymerases like Phusion Hot Start II High-Fidelity DNA Polymerase (Biozym) for avoiding or minimizing stutter bands. We have obtained good results with these two polymerases.

14. Allele-specific primer design is carried out with IDT PrimerQuest® program [30]. The last base at the 3′ end of one primer (forward or reverse) confers the allele-specificity (*see* Fig. 8); whereas the second primer is able to amplify both

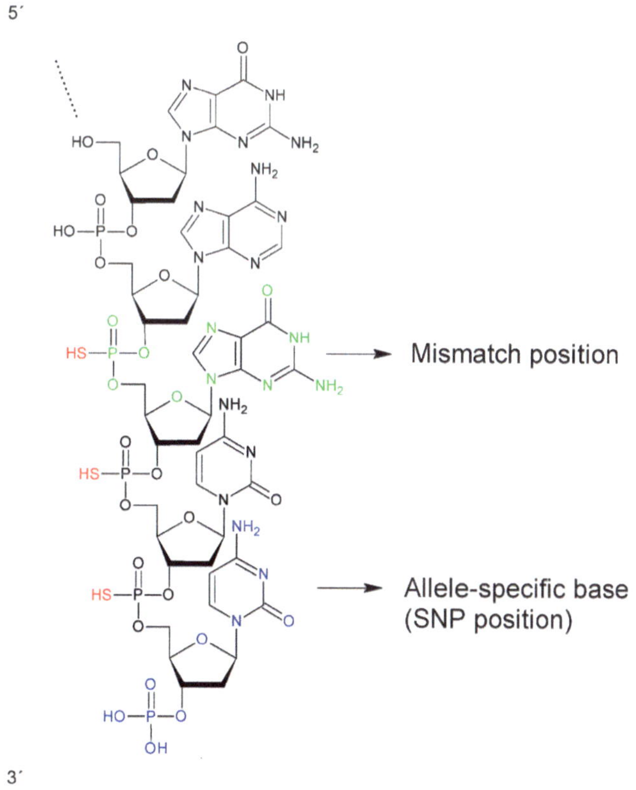

Fig. 8 Allele-specific primer modifications. This figure shows a forward primer in 5′–3′ direction. The phosphate backbone of base 2–4 is modified with sulfur atoms (*red*) to avoid the degradation of the allele-specific base by the 3′–5′ proof-reading activity of the polymerase. Additionally, phosphorothioate (PTO) modifications make the primer more rigid, so that a single nucleotide mismatch has a strong influence on the binding between primer and template. At the third position from the 3′ end, an additional mismatch (*green*) can be introduced to increase the sequence specificity. The 3′ base (*blue*) represents the allele-specific base

alleles. For example, three primers have to be designed for a C>T SNP, two different allele-specific forward primers with the base at the 3′ end complementary to either C or T and one universal reverse primer, ending up in the primer combinations *primer-C + universal* and *primer-T + universal*. In order to increase the specificity of the primers, phosphorothioate (PTO) bonds are introduced into the backbone of the last four bases at the 3′ end. Additionally, a second mismatch at the third position from the 3′ end can increase the reaction specificity.

Allele-specific genotyping can also be used to type a STR length polymorphism (here we show an example of an STR with 19A or 9A; *see* Table 8, primer set 3). This can be achieved with allele-specific primers that include mismatches, tails to increase the GC content, and PTO bonds. For genotyping this STR, we obtained a better specificity when using up to four instead of just one allele-specific base at the 3′ end placed outside the STR instead. The PCR conditions are adapted for a low GC content and differ significantly from the normal Phusion Hot Start II

Table 8

Allele-specific genotyping assay design

	Allele-specific genotyping assay design					
	Sequence	**Primer length [bp]**	**T_m [°C]**	**GC content [%]**	**Product length [bp]**	**Optimized annealing temp [°C]**
Primer set 1: SNP C/T—OneTaq Hot Start DNA Polymerase (*see* Table 5)						
SNP rs1861187 C/T (genome position: 6.359.077, GRCh37/hg19)						
Forward primer 1 (C, allele-specific)	GCGATTGAAATAATCAGGTCg* C*A*C	24	59.3	59	69	63
Forward primer 2 (T, allele-specific)	GCGATTGAAATAATCAGGTCg* C*A*T	24	57.6	58		
Reverse primer	GAATTCAAAACAGGCGAACG	20	55.3	45		
Primer set 2: STR length polymorphisms 7A/6A—Phusion Hot Start II High-Fidelity DNA Polymerase (*see* Table 6)						
rs35094442 7A/6A (genome position: 6.310.566, GRCh37/hg19)						

(continued)

Table 8 (continued)

	Allele-specific genotyping assay design					
	Sequence	Primer length [bp]	T_m [°C]	GC content [%]	Product length [bp]	Optimized annealing temp [°C]
Forward primer	GCTGTAGTGTCCTCACATCAACCC	24	64.4	54	82/81	60
Reverse primer 1 (7A, allele-specific)	CCGCTTGGAGCTTCAGTTTT*g*T*T	23	60.6	48		
Reverse primer 2 (6A or C, allele-specific)	CCGCTTGGAGCTTCAGTTTT*g*T*G	23	62.4	52		
Primer set 3: STR length polymorphisms 19A/9A—Phusion Hot Start II High-Fidelity DNA Polymerase (*see* Table 9)						
rs200121160 19A/9A (genome position: 6.360.903, GRCh37/hg19)						
Forward primer 1 (19A. allele-specific)	GCCGCACATTTACCAGTGGTTTAAAAAAtAAA*A*A*A	35	63.6	31	94/107	56
Forward primer 2 (9A. allele-specific)	GCACATTTACCAGTGGTTTAAAAAAtAAG*A*A*C	32	61.8	31		
Reverse primer	TGTCCTAGCATCTCTGATAAC	21	55.9	43		

Shown are allele-specific primer pairs for three different polymorphisms on the human chromosome 16. The nucleotide at the 3′ end base is the allele-specific base (bold and underlined). The bases marked with a star (*) represents nucleotides connected with a PTO bond; additional mismatches are shown in lower case. In some cases, a 5′ tail, which is not included in the sequence (italic and underlined), is added to increase the T_m or GC content of the primer. The two last primer sets are specific for genotyping STRs length polymorphism (7A/6A and 19A/9A). In contrast to SNP genotyping, for the longer STR (19A/9A), the allele-specific primer included more than one allele-specific site at the 3′ end (4 bases outside the repeat) in order to avoid primer misalignments, and an additional mismatch in the middle to break-up the long run of poly As.

High Fidelity program; although, the polymerase is the same. It is important to note that these primers need a strict optimized PCR temperature profile. Details are explained further down in **Note 35**. Table 8 shows allele-specific primers and PCR conditions for our region on chromosome 16.

15. For allele-specific genotyping, intercalating fluorescent dyes are required. We recommend SYBR Green I or EvaGreen, if premixed mastermixes are not used. EvaGreen is a saturating dye resulting in a higher sensitivity than non-saturating SYBR Green I. Additionally, EvaGreen is less inhibitory to PCR than SYBR Green I, which could be a problem in very difficult DNA templates [36, 37]. Note that SYBR Green I is dissolved in DMSO. DMSO increases the specificity of an allele-specific assay by binding to cytosine residues, and decreasing the melting temperature of GC-rich regions. It also facilitates the correct primer annealing to the template. In case of a mismatch of the allele-specific nucleotide, the binding of the perfect matching primer is more stable than that of mispaired primers.

16. TaqMan assay optimization is a very critical process, carried out in two steps. In the first optimization step, only the primers flanking the polymorphism are optimized to ensure high yields of the expected product and minimize unspecific product formation. Reactions producing multiple nonspecific products, observed as numerous bands in gel electrophoresis, often render ambiguous genotyping results and should be avoided. In the second optimization step, primer and probes are optimized together, but this step concentrates testing the allelic distinction of the probes. Separating the optimization steps of primer and probes has several advantages: (1) it is easier to troubleshoot when a reaction generates unspecific signals, wrong genotype calls or does not work at all; (2) a specific faulty step can be traced back to the flanking primers or the probes; and (3) the probe specificity and efficiency can be evaluated better (e.g. does one probe perform better than the other. In the long run it is less laborious to carry out two optimization steps than trying to troubleshoot a combination of different factors.

    For the *first optimization step*, the TaqMan PCR protocol is carried out with a temperature gradient in the annealing step of the PCR cycle in a reaction *without the probes*. The gradient is chosen in such a way that the lowest temperature matches the predicted melting temperature of the primer (usually stated by the oligo synthesis company or by the software used to design the primers), up to ~T_m +9 °C for three to four different temperature steps (e.g. 56–59–62–65 °C). This temperature range should also include the T_m of the probe, since the T_m of the probes should be 5–10 °C higher than of the primers (*see* **Note 11**). In this optimization step, we use an intercalating fluorescent dye like SYBR Green I or

EvaGreen to visualize the amplicons in rtPCR and also perform a melting curve analysis after the cycling steps. Melting curves are quite useful to infer the presence of nonspecific products (present as extra peaks). We still recommend to visually inspect the sizes of the produced amplicons in a 10% DNA-polyacrylamide gel (DNA-PAA gel) or a high-resolution agarose gel.

In the *second optimization step*, the dual-labeled probes are added, but in this case the T_m of the TaqMan probes is used as the lowest temperature setting of a temperature range up to ~T_m +6 °C. Do not use an additional fluorescent dye or a melting curve analysis step when setting up the program of the thermocycler. Choose several DNA templates with known genotypes (heterozygotes and homozygotes, as well as non-template controls). This will help to evaluate the genotyping accuracy of the probe. The optimal temperature within the gradient is the temperature that renders the correct genotyping, a strong signal measured as relative florescence units (RFU), as well as, the absence of unspecific products or signals.

If the signal intensity is too low, try different annealing/extension temperatures and adjust the $MgCl_2$ concentration in 0.5–1 mM steps. Please note that lowering the annealing/extension steps enhances the binding stability for AT-richer TaqMan probes and therefore increases the signal intensity, but at the same time reduces the binding specificity of the GC-richer probe leading to lower fluorescence signals. This is also true for the reverse case, when the annealing/extension step is increased, facilitating the binding of the GC-richer TaqMan probe.

17. Before preparing the mastermix or working dilutions, mix all the stock solutions after thawing to properly dissolve the salts, either by flicking or by vortexing followed by a quick spin centrifugation step. Otherwise it is possible that the concentration of the stock solutions varies, leading to wrong signals.
18. Prepare the TaqMan mastermix in an appropriate-sized tube. For example, a 384-well plate (425× mastermix) needs 10 μl per reaction resulting in 4250 μl total mastermix. Prepare 10% more of the total mastermix volume representing waste volumes that account for pipetting errors.
19. Store the polymerase at −20 °C or on ice until use. Pipette out the required volume for the mastermix and refreeze the remaining stock solution immediately.
20. Mastermixes can be stored for several days at 4 °C under light exclusion. A decreased signal intensity of the probes has not been observed.
21. Given the small volumes and the large numbers of reactions, set up the mastermix on ice and also pipette out the plates on ice to avoid evaporation of the mastermix. We use multichannel

pipetting basins to rapidly aliquot the mastermix into the 96- or 384-well plates.

22. We use multichannel pipettes (1–10 μl or 2–20 μl) to add the DNA templates from a 384-well plate into the mastermix plate. This ensures a fast and clean method for adding the DNA.
23. Close the PCR plates carefully with rtPCR suitable seals. You can use a PCR seal hand applicator or something else suitable like a piece of thick plastic for smoothing the foil on the plate. We do not use a heat plate sealer. Seal the borders carefully. Seals that are not properly glued on, or with wrinkles, can distort the rtPCR signal due to evaporation of the reaction fluid.
24. After sealing the rtPCR plate, mix the mastermix with the DNA properly by turning and tapping the plate up-side down for several times and followed by a quick spin-down (a touch spin is sufficient for 96-well plates, but for 384-well plates we recommended a centrifugation step for 2 min at 2000 × *g*). This is an important step to get rid of air bubbles in the reaction wells that could lead to high fluorescence background or false signals.
25. PCR plates can be stored in a plastic bag with a wet piece of kitchen roll to avoid further evaporation and loss of water, which would change the concentrations of the reagents in the mastermix. We have not observed a decrease in fluorescence intensity when the plate has been stored for 1–2 days in the fridge at 4 °C before the PCR has been run.
26. The PCR cycling parameters will depend on the polymerase. Normally, these parameters can be found in the product information or specification sheets.
27. Our PCR program has been optimized in several respects. We recommend denaturation steps at lower temperatures and shorter times to avoid DNA damage. Additionally, we choose short combined annealing/extension steps for 70–150 bp products to avoid nonspecific products. Polymerases tend to choose "any" template, if the amplification time is too long. Longer templates require longer annealing/extension times. Optimizing the cycling parameters is critical for a well-working genotyping assay.
28. Commercial available rtPCR cyclers have preinstalled settings for the most common dyes. Before an assay is designed, check if the dyes are calibrated for your instrument which you would like to use. Before the rtPCR is started, choose the right fluorophore for the plate read. Otherwise the signals will be wrong or absent.
29. On several occasions we have observed that one fluorophore renders a higher signal than the other, likely due to chemical differences in the fluorophores affecting the light emission or due to the fact that one probe binds more stable than the other. For those cases, we switch the 5′ dyes of the probes. That means

that we order new probes, but now with the opposite labeling ending up with equal RFUs between probes (*see* Fig. 9)

30. Figure 5a in Subheading 3.1, **step 5** shows amplification curves obtained in the lab. Some of them show a lower RFU value at the end phase. These are heterozygote samples that contain half of the effective template number compared to homozygous samples. Nevertheless, well optimized probes should show roughly the same fluorescence intensities (RFUs) for both probe moieties.

31. Allele-specific primer optimization is quite similar to the TaqMan primer optimization. Use a temperature gradient for the annealing step of the PCR cycle, with the T_m of the primers, T_m +3 °C, T_m +6 °C and T_m +9 °C (e.g. 60–63–66–69 °C). The products are separated on a gel (e.g. 10% DNA-PAA gel). If there are multiple peaks in the melting curve analysis or several additional bands in the gel electrophoresis, try to optimize the reaction with different cycling temperatures/times, $MgCl_2$

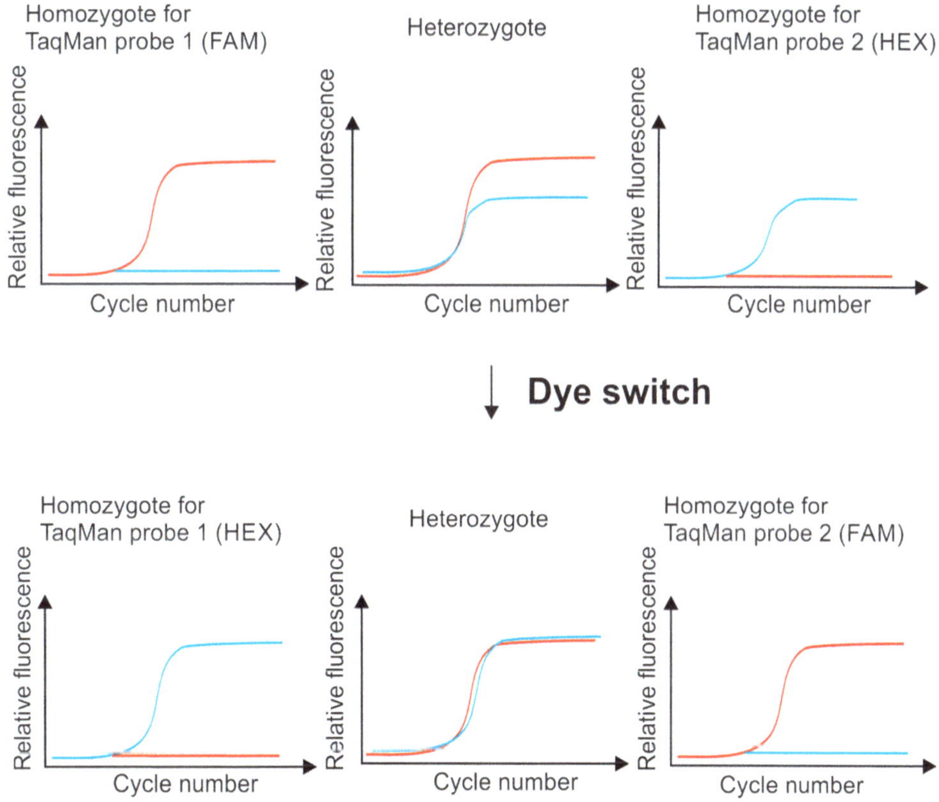

Fig. 9 Dye switch of dual-labeled probes. In some cases one dual-labeled probe results in a higher final relative fluorescence signal (RFU) than the counter probe, resulting in problems with the genotype call. This could be due to a stronger annealing of one probe to the template (higher GC content or T_m) and if this probe is additionally labeled with a dye with a lower fluorescence intensity due to dye chemistry, the RFU values are reduced. By switching the dyes of the probes without changing the sequence, the problem can often be solved

concentration up to 3 mM, DMSO concentration up to 10% or design new primers. The genotype is defined by the reaction that produced an earlier amplification curve of the two reactions, each containing the same DNA template but a different primer combination. There should be at least ~5 cycles difference between the two reactions of a homozygous sample, and they should have nearly the same inflection points for heterozygote samples.

32. Note that in a 384-well plate only 192 DNA samples can be genotyped, because two reactions per sample are needed. For a 384-well plate prepare a 465× mastermix, sufficient for two 211× mastermixes for primer combination 1 and 2 (10% more for 192 reactions and 10% more for the whole mastermix).

33. We use multichannel pipetting basins to easily aliquot the mastermix into the well plates. The mastermix can be divided into two tubes or directly into the multichannel pipetting basins. Aliquot the 211× mastermix into a basin and then add the allele-specific primer combination 1 in basin 1 and the allele-specific primer combination 2 in basin 2. By slewing the basins several times, the mastermix and primer solutions are mixed. Using directly the basins and no tubes, pipetting errors are avoided which can lead to different primer concentrations. For the easier analysis of allele-specific genotyping reactions, use the odd wells for one allele- and the even wells for the alternate allele-specific reaction. Do not pipette the primer combinations that you wish to compare into separate plates! You can even use the same pipette tip for adding the DNA to both primer combinations for one sample. The conditions during a PCR program have to be identical. Also slight differences in pipetting can make pivotal differences in separation between the genotypes.

34. We recommend the use of larger DNA volumes with lower concentrations to avoid pipetting errors which can result in wrong genotypes. For example, use 5 μl of a 2 ng/μl DNA stock solution instead of 1 μl of a 10 ng/μl stock solution (final concentration 10 ng human genomic DNA or 1500–3000 molecules).

35. The temperature program is unique for each polymerase, and the annealing steps are optimized to yield a high amplification and the correct genotype of the DNA template. We use lower denaturation temperatures (e.g. 94 °C/95 °C instead of 98 °C recommended in the vendors manual) to reduce biases due to DNA lesions generated at high temperatures. The annealing and extension steps are rather short for Phusion Hot Start II High-Fidelity polymerase (5 s annealing and 10 s extension) to avoid nonspecific products. If the amplification time is too long, polymerases tend to amplify more nonspecific products.

Longer templates require longer annealing/extension times. When we genotype STR length polymorphisms with more than 10 consecutive A/Ts, we use Phusion Hot Start II High-Fidelity DNA Polymerase and a slightly different PCR program, as shown in Table 9. The extension step of the OneTaq Hot Start DNA Polymerase program is slightly different from the Phusion Hot Start II High-Fidelity Polymerase program. This is due to the optimization for unique primer pairs and can differ in other applications.

36. After ~ 30–45 PCR cycles followed by a final extension step, a melt curve analysis is performed. By increasing the temperature slowly in 0.5 °C steps, the florescence decreases given that the DNA denaturation releases the intercalating dye. The change in fluorescence signal per temperature is used to plot the melting curve. The maximum of the melting curve inflection represents the melting temperature of the PCR product. The melting peak plot can be created by plotting the change in fluorescence signal per temperature ($-\Delta F/\Delta T$) against temperature, resulting in sharp peaks for each PCR product.
37. If the separation between the two amplification curves for homozygote samples is still smaller than 5 cycles after the optimization steps (increased DMSO concentrations, optimized temperature protocol, increased $MgCl_2$ concentration), we recommend designing new primers.

Table 9

rtPCR cycling protocol for allele-specific genotyping assays with Phusion Hot Start II High-Fidelity DNA Polymerase for sequences with STRs with more than 10 consecutive As or Ts

Allele-specific PCR cycling protocol for STR length polymorphism >10 consecutive A/Ts			
Step	**Temperature**	**Time**	**Note**
1 activation	94 °C	2 min	
2 denaturation	94 °C	15 s	Go to **step 2**; repeat 5×
3 annealing	T_m	5 s	
4 extension	63 °C	15 s	
5 denaturation	94 °C	15 s	Go to **step 2**; repeat 40×
6 annealing	53 °C	5 s	
7 extension	58 °C	15 s	
	Plate read for SYBR Green I		
8 final elongation	58 °C	30 min	
9 melting curve	65–95 °C	0.5 °C/min	

References

1. Mullis KB, Faloona FA (1987) Specific synthesis of DNA in vitro via a polymerase-catalyzed chain reaction. Methods Enzymol 155:335–350
2. Higuchi R, Dollinger G, Walsh PS et al (1992) Simultaneous amplification and detection of specific DNA sequences. Biotechnology (NY) 10:413–417
3. Higuchi R, Fockler C, Dollinger G et al (1993) Kinetic PCR analysis: real-time monitoring of DNA amplification reactions. Biotechnology (NY) 11:1026–1030
4. Saiki RK, Walsh PS, Levenson CH et al (1989) Genetic analysis of amplified DNA with immobilized sequence-specific oligonucleotide probes. Proc Natl Acad Sci U S A 86:6230–6234
5. Southern EM (1975) Detection of specific sequences among DNA fragments separated by Gel-electrophoresis. J Mol Biol 98:503–517
6. Holland PM, Abramson RD et al (1991) Detection of specific polymerase chain reaction product by utilizing the 5′-3′ exonuclease activity of Thermus aquaticus DNA polymerase. Proc Natl Acad Sci U S A 88:7276–7280
7. Ririe KM, Rasmussen RP, Wittwer CT (1997) Product differentiation by analysis of DNA melting curves during the polymerase chain reaction. Anal Biochem 245:154–160
8. VanGuilder HD, Vrana KE, Freeman WM (2008) Twenty-five years of quantitative PCR for gene expression analysis. Biotechniques 44:619–626
9. Navarro E, Serrano-Heras G, Castano MJ et al (2015) Real-time PCR detection chemistry. Clin Chim Acta 439:231–250
10. Wittwer CT, Herrmann MG, Moss AA et al (1997) Continuous fluorescence monitoring of rapid cycle DNA amplification. Biotechniques 22(130–1):34–38
11. Chou Q, Russell M, Birch DE et al (1992) Prevention of pre-PCR mis-priming and primer dimerization improves low-copy-number amplifications. Nucleic Acids Res 20:1717–1723
12. Cardullo RA, Agrawal S, Flores C et al (1988) Detection of nucleic acid hybridization by non-radiative fluorescence resonance energy transfer. Proc Natl Acad Sci U S A 85:8790–8794
13. Forster T (1948) Zwischenmolekulare Energiewanderung Und Fluoreszenz. Ann Phys-Berlin 2:55–75
14. Cobos-Correa A, Schultz C (2009) Small molecule-based FRET probes. In: Gadella TWJ (ed) Laboratory techniques in biochemistry and molecular biology, vol 33. Academic Press, Heidelberg, pp 225–288
15. Sekar RB, Periasamy A (2003) Fluorescence resonance energy transfer (FRET) microscopy imaging of live cell protein localizations. J Cell Biol 160:629–633
16. Wang JC (1979) Helical repeat of DNA in solution. Proc Natl Acad Sci U S A 76:200–203
17. Whitcombe D, Theaker J, Guy SP et al (1999) Detection of PCR products using self-probing amplicons and fluorescence. Nat Biotechnol 17:804–807
18. Nazarenko I, Lowe B, Darfler M et al (2002) Multiplex quantitative PCR using self-quenched primers labeled with a single fluorophore. Nucleic Acids Res 30, e37
19. Kandimalla ER, Agrawal S (2000) 'Cyclicons' as hybridization-based fluorescent primer-probes: synthesis, properties and application in real-time PCR. Bioorg Med Chem 8:1911–1916
20. Lee MA, Siddle AL, Page RH (2002) ResonSense (R): simple linear fluorescent probes for quantitative homogeneous rapid polymerase chain reaction. Anal Chim Acta 457:61–70
21. Wittwer CT, Ririe KM, Andrew RV et al (1997) The LightCycler: a microvolume multisample fluorimeter with rapid temperature control. Biotechniques 22:176–181
22. Kumar R, Singh SK, Koshkin AA et al (1998) The first analogues of LNA (locked nucleic acids): phosphorothioate-LNA and 2′-thio-LNA. Bioorg Med Chem Lett 8:2219–2222
23. Nielsen PE, Egholm M, Berg RH et al (1991) Sequence-selective recognition of DNA by strand displacement with a thymine-substituted polyamide. Science 254:1497–1500
24. de Noronha CM, Mullins JI (1992) Amplimers with 3′-terminal phosphorothioate linkages resist degradation by vent polymerase and reduce Taq polymerase mispriming. PCR Methods Appl 2:131–136
25. Vester B, Wengel J (2004) LNA (locked nucleic acid): high-affinity targeting of complementary RNA and DNA. Biochemistry 43:13233–13241
26. Singh SK, Nielsen P, Koshkin AA et al (1998) LNA (locked nucleic acids): synthesis and high-affinity nucleic acid recognition. Chem Commun 4:455–456
27. Lee LG, Connell CR, Bloch W (1993) Allelic discrimination by nick-translation PCR with fluorogenic probes. Nucleic Acids Res 21:3761–3766
28. Livak KJ, Flood SJ, Marmaro J et al (1995) Oligonucleotides with fluorescent dyes at opposite ends provide a quenched probe

system useful for detecting PCR product and nucleic acid hybridization. PCR Methods Appl 4:357–362
29. Tiemann-Boege I, Calabrese P, Cochran DM et al (2006) High-resolution recombination patterns in a region of human chromosome 21 measured by sperm typing. PLoS Genet 2, e70
30. Western PS, Surani MA (2002) Nuclear reprogramming--alchemy or analysis?[comment]. Nat Biotechnol 20:445–446
31. Untergasser A, Nijveen H, Rao X et al (2007) Primer3Plus, an enhanced web interface to Primer3. Nucleic Acids Res 35:W71–W74
32. Bio-Rad Laboratories Inc. Real Time PCR Application Guide [http://www.gene-quantification.de/real-time-pcr-guide-bio-rad.pdf]
33. Reynisson E, Josefsen MH, Krause M et al (2006) Evaluation of probe chemistries and platforms to improve the detection limit of real-time PCR. J Microbiol Methods 66:206–216
34. Huang Q, Zheng L, Zhu Y et al (2011) Multicolor combinatorial probe coding for real-time PCR. PLoS One 6, e16033
35. Chevalier A, Hardouin J, Renard PY et al (2013) Universal dark quencher based on "clicked" spectrally distinct azo dyes. Org Lett 15:6082–6085
36. Mao F, Leung WY, Xin X (2007) Characterization of EvaGreen and the implication of its physicochemical properties for qPCR applications. BMC Biotechnol 7:76
37. Monis PT, Giglio S, Saint CP (2005) Comparison of SYTO9 and SYBR Green I for real-time polymerase chain reaction and investigation of the effect of dye concentration on amplification and DNA melting curve analysis. Anal Biochem 340:24–34

Chapter 4

In Situ Single-Molecule RNA Genotyping Using Padlock Probes and Rolling Circle Amplification

Tomasz Krzywkowski, Thomas Hauling, and Mats Nilsson

Abstract

Present-day techniques allow for massively parallel and high-throughput characterization of the somatic mutation status of samples. Most of these assays rely on whole specimen extracts, where heterogeneous spatial context of the specimen is lost. This chapter describes an up-to-date protocol for multiplexed, in situ genotyping of RNA in preserved tissue and cell lines, using padlock probes and rolling circle amplification. The presented approach allows for automated quantification of mRNA expression and mutation status, in single cells or in designated specimen areas. Briefly, mRNA is first reverse-transcribed to cDNA. Padlock probes specifically hybridize to the cDNA copy of the allele and become circularized and thereby physically linked to their targets. Following this conversion, padlock probes are copied in situ by rolling circle amplification and labeled with flourophore-conjugated probes, allowing for their detection with conventional fluorescence microscopy.

Key words Padlock probe, mRNA genotyping, In situ, Single cell

1 Introduction

Controlled expression of genes is a central metabolic measure that underlies cell development, homeostasis, and death. Gene sequence alteration, whether as an effect of imprecise DNA replication or deleterious environmental conditions, can lead to defective cell response or promote tumor growth [1, 2]. Defining spatial localization of mutations in a specimen can aid in the interpretation of sample complexity, better understand disease processes or guide therapeutical predictions. Over the years, multiple techniques have emerged to support precise in situ quantification of DNA or mRNA, in addition to traditional fluorescent RNA in situ hybridization (FISH) methods. These include: single-molecule (sm) FISH (based on hybridization of multiple labeled fluorescent probes along a target RNA strand) [3]. Optionally, combinations of smFISH probes can be used to determine mRNA identity [4]. Single molecule resolution can also be achieved by branched DNA (bDNA) FISH [5]; and

Stefan J. White and Stuart Cantsilieris (eds.), *Genotyping: Methods and Protocols*, Methods in Molecular Biology, vol. 1492, DOI 10.1007/978-1-4939-6442-0_4, © Springer Science+Business Media New York 2017

hybridization chain reaction (HCR) [6]. All three methods rely on hybridization of multiple probes (at least two, approach-dependent) in close proximity along a target RNA strand to generate a scaffold for signal amplification. Hence, specificity is a function of interrogated strand length. Therefore, genotyping that requires querying of single nucleotide variations (SNVs) remains challenging with methods that solely rely on hybridization.

Here, we present an updated protocol and set of guidelines on how to genotype single mRNA molecules in situ, using padlock probes and rolling circle amplification (RCA). Padlock probes are linear, single-stranded DNA oligonucleotides composed of two target-complementary termini and a linker segment [7]. In a typical padlock probe-based assay, two probe arms hybridize to the target juxtaposed, and a nick is sealed by a mismatch-sensitive DNA ligase. Compared to FISH methods, padlock probes, supported by enzymatic ligation accuracy, offer superior discrimination specificity, allowing for detection of SNVs [8]. For genotyping by padlock probing, mRNA molecules are reverse transcribed to cDNA (*see* **Note 1**). After target mRNA is degraded by ribonuclease H (RnaseH explicitly degrades RNA from RNA/cDNA heteroduplexes), allele-specific padlock probes are hybridized. Depending on the variant present, corresponding padlock probes will become ligated by *Tth* DNA ligase and thereby circularized (*see* **Note 2**). It is the ligation step that confers SNV specificity to padlock probing, due to the sensitivity of *Tth* ligase for mismatches at the ligation site.

Complete DNA circles, concatenated with their targets, serve as a template for ϕ29 polymerase-driven amplification [9]. This step generates continuous, single-stranded DNA products comprising 10^2–10^3 tandem copies of the original padlock probes [10]. Since, in the described method, each target cDNA serves as a primer with a free 3′ –OH group, the amplified rolling circle product (RCP) remains physically bound to its target mRNA (*see* Fig. 1).

RCPs spontaneously coil into spherical DNA "balls" (500 μm diameter on average) and can be visualized by hybridizing fluorophore-conjugated oligonucleotides (decorator probes), complementary to motifs in the probe amplicons. RCPs are readily differentiated from background since they contain hundreds of decorator probe hybridization sites. Padlock probes can be designed with unique linker sequences to allow for simultaneous detection of multiple targets using linker-specific decorator probes (conjugated with different fluorophores). Genotyping by padlock probing allows for exact quantification of variants since each RCP corresponds to a single mRNA molecule. Owing to the discrete shape and typically sub-micron size of RCPs, transcripts can be mapped with

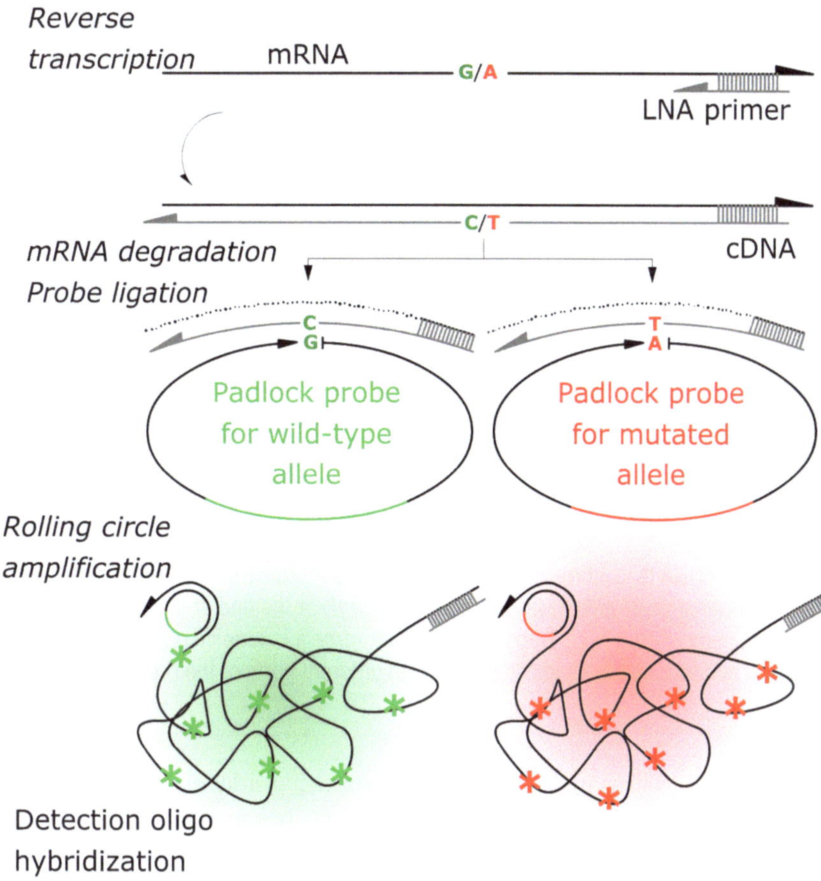

Fig. 1 Outline of in situ RNA genotyping using padlock probes and target primed rolling circle amplification. An mRNA molecule harboring a sequence variant is converted into cDNA by reverse transcription, using target-specific (ideally LNA-modified) primers or random decamers. Subsequently, the mRNA is degraded using RnaseH, padlock probes for the respective alleles hybridize to the target sequence and circularize upon DNA ligation. RCA generates a single-stranded DNA concatamer, which collapses into a typically micrometer-sized DNA ball and contains hundreds of tandem repeated sequences that are complementary to the original padlock probe. Finally, fluorescently labeled decorator probes hybridize with their complementary motifs on the RCA product

subcellular resolution. We apply the presented procedure to detect somatic SNVs in cell lines and tissue.

2 Materials

2.1 Oligonucleotides

2.1.1 Primers

We recommend two strategies for primer design. For highly abundant mRNAs, random degenerate primers often achieve satisfactory sensitivity (*see* **Note 3**). For mRNAs expressed at low levels, we advise design of mRNA-specific LNA-modified primers, to maximize efficiency of reverse transcription. Such primers are designed

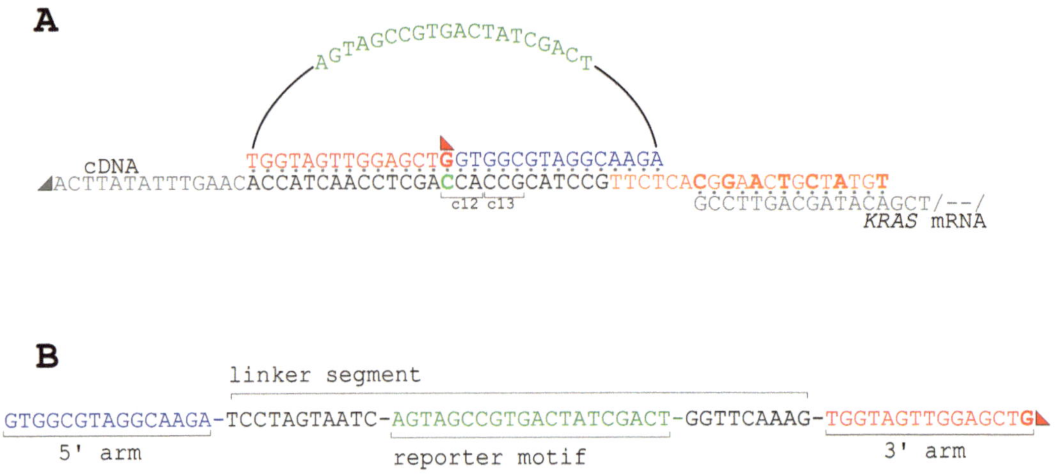

Fig. 2 Padlock probe and LNA-primer design blueprint for *KRAS* codon 12 and 13. (**a**) LNA primer (*orange*; LNA bases: *bold*) hybridizes with *KRAS* mRNA. During the ligation, mRNA is degraded except where LNA bases were introduced. This "locks" cDNA to its corresponding mRNA. Padlock arms (*red* and *blue highlight*) hybridize with the target, while discriminative 3′ base (symbolized as a *triangle* at the end of the *red* 3′ arm) is located over the interrogated, first base of codon 12. Reporter sequence (*green*) is amplified and used later for detection. (**b**) 5′→3′ full-length sequence of the padlock probe, with different parts of the probe indicated

following standard PCR primer guidelines, i.e. they are about 25 nucleotides long. Five to seven bases are typically substituted with chemically altered, LNA nucleotides (*see* **Note 4**). Primers should be positioned 20–100 bp upstream from the mutated site (*see* **Note 5**). LNA bases should be excluded from target-overlapping sites since they can inhibit amplification of hybridized padlock probes. Using *c.12 KRAS* mutation detection as an example, Fig. 2 illustrates a typical design strategy for padlock probe and corresponding primer.

2.1.2 Padlock Probes and Decorator Oligonucleotides

We advise using the open-source software ProbeMaker [11] for automated padlock probe design (*see* **Note 6**). Terminal arms are designed to base pair with the target site, harboring the mutation. Melting point (Tm) of the padlock probe arms (and thus, length) should be adjusted for the ligation step conditions (*see* **Note 7**). Arm length may vary, as Tm depends mostly on the GC content of the target sequence. Increasing arm length to strengthen the hybridization should be avoided since this increases the risk that even a partially complementary probe for the "wrong" allelic variant would block availability of the detection site. If multiple probes are used in the experiment, the Tm of all probes should be similar to ensure comparable probe performance. A discriminating nucleotide of the padlock probe (one hybridizing with a mutated equivalent) should be located at 3′ end of the probe, as such a design maximizes ligation specificity [12]. A linker segment harboring a reporter motif (unique for each allele-specific probe) is placed between padlock probe arms, and should be 10 nt longer than the target sequence (*see* **Note 8**). After the padlock probe is amplified, the product will contain multiple, complementary repeats of the original probe, including the reporter motif. We take advantage of these unique motifs to differentiate RCPs (*see* **Note 9** for decorator probe design guidelines). After the full-length padlock probe sequence is determined, we advise prediction of the probe secondary structure. We use mfold (http://mfold.rna.albany.edu [13]) or OligoAnalyzer as they provide intuitive engines, where multiple parameters concerning the hybridization conditions can be set. To enable ligation, padlock probes must have a 5′ phosphate group. In our hands, pre 5′-phosphorylated oligonucleotides, synthesized by IDT (http://idtdna.com) using Ultramer® chemistry work well. More design guidelines and a phosphorylation protocol can be found in **Note 10**. For a list of all oligonucleotides used in this chapter, *see* Table 1.

2.2 Reagents

All enzymes should be stored at −20 °C. Other reagents are stored at room temperature (RT) unless stated different.

1. RIBOPROTECT—RNase Inhibitor 40 U/μl (DNA Gdansk).
2. TRANSCRIPTME—reverse transcriptase 200 U/μl and buffer (*see* **Note 11**).
3. *Tth* DNA ligase 40 U/μl.

Table 1

Padlock probes and primer sequences used in the present chapter

Primers	Sequences (5′–3′)
KRAS-c12/13[b]	T+GT+AT+CG+TC+AA+GG+CACTCTT
Padlock probes	Sequences (5′–3′)
KRAS-wt[a]	GTGGCGTAGGCAAGATCCTAGTAATC *AGTAGCCGTGACTATCGACT* GGTTCAAAG TGGTAGTTGGAGCTG
KRAS-G12S[a]	GTGGCGTAGGCAAGATTCTAGATC *CCTCAATGCACATGTTTGGCTCC* GGTTCAAG TGGTAGTTGGAGCTA
Detection probes	Sequences (5′–3′)
KRAS-wt[a]	AGTAGCCGTGACTATCGACT
KRAS-G12S[a]	CCTCAATGCACATGTTTGGCTCC

+, LNA-modified base; *underline*, target complementary arms; *italic*, detection probe complementary sequence. Oligonucleotides were purchased from: Integrated DNA Technologies[a], Exiqon[b]

4. RNaseH 5 U/µl.
5. Phi29 DNA polymerase 10 U/µl and buffer.
6. T4 PNK Kinase and buffer
7. ATP 100 mM solution. Stored at −20 °C.
8. BSA 20 mg/ml. Stored at −20 °C.
9. Biological specimen: cultured cells (alive) or tissue (fresh or fresh frozen) of interest (*see* **Note 12**).
10. Diethylpyrocarbonate (DEPC). Stored at 4 °C (*see* **Note 13**).
11. RNase AWAY™ (Invitrogen) and DNase Away (Genemark).
12. dNTP set of 100 mM solutions. Stored at −20 °C.
13. Ethanol (70, 85, 99.5 %, v/v).
14. Formamide (*see* **Note 14**).
15. Glycerol.
16. Hydrochloric acid (*see* **Note 15**).
17. Formaldehyde (*see* **Note 16**).
18. Pepsin lyophilized powder 2500 U/mg protein (Sigma-Aldrich) (*see* **Note 17**).
19. Potassium chloride.
20. Trypsin-EDTA 0.25 %.

    Used in fixation of adherent cells.
21. SlowFade® Gold Antifade Mountant (Thermo Scientific) or equivalent mounting medium.
22. Hoechst 33342 (Thermo Scientific)

    Stock Hoechst 33342 solution should be kept at −20 °C. *Working solutions can be stored at 4°C for a couple of months.*

2.3 Solutions and Buffers

Concentrated buffers are provided with enzymes by the respective vendors and are stored according to specification. Custom-made buffers should be prepared from DEPC-treated PBS or ddH_2O (*see* **Note 18**) and can be kept at RT.

1. RT—reverse transcriptase buffer (10×).
2. Tth DNA ligase buffer (10×).
3. Phi29 DNA polymerase buffer (10×).
4. Phosphate buffered saline 1× PBS pH 7.4: NaCl 137 mM, sodium phosphate 10 mM, KCl 2.7 mM, DEPC-ddH_2O.
5. Washing buffer 1× DEPC-PBS-T pH 7.4: 0.05 % Tween 20, 1× DEPC-PBS.
6. Saline-sodium citrate buffer 20× SSC pH 7: NaCl 3 M, trisodium citrate 300 mM, DEPC-ddH_2O.
7. 2× Hybridization mix: SSC 4×, formamide 40 % (v/v))

Store at RT, protected from light.

2.4 Equipment

1. Diamond pen.
2. Forceps.
3. Incubator 37 °C.
4. Incubator 45 °C.
5. Humidity chamber (e.g. empty tip box with water-soaked paper towel in the bottom).
6. Fluorescence microscope.
7. Secure-Seal chambers-
   Size depends on the experiment setup (see **Note 19**).
8. Cover glasses.
   To achieve optimal optical resolution, the type of cover glass needs to be adjusted for the desired microscope setup.
9. SuperFrost® microscopy slides (Menzel Gläser).
10. 150 mm × 25 mm culture dish (Corning).

3 Methods

3.1 General Recommendations and Controls

All consumables (gloves, filtered tips, etc.) should be RNase-free. Reaction mixtures should be prepared from DEPC-ddH_2O or DEPC-PBS. We recommend cleaning lab benches and reusable labware from RNases, RNA or DNA traces with chemical reagents.

We recommend validating specificity of padlock probes on synthetic DNA oligonucleotides that span the target region. Ligation can be performed in vitro and monitored as a high-molecular-weight band on denaturing PAGE gel (unligated probes and target oligonucleotides will migrate faster than circularized probes). Alternatively, ligated padlock probes can be amplified in vitro (short target provides the 3′ –OH group as a primer), stained with DNA intercalating dyes or complementary flourophore-conjugated decorator probes, and visualized under a microscope. Cell lines with known expression data for targets of interest provide a good model to assess detection specificity and efficiency of padlock probes in biological specimens.

3.2 Sample Preparation

3.2.1 Adherent Cell Lines

1. Cells are cultured in flask until confluent.
2. Wash cells twice with 1× PBS, and treat with 0.25% (w/v) Trypsin-EDTA.
3. Resuspend cells in appropriate culturing medium.
4. Place slides in a petri dish and add ~22 ml of medium to cover the slides (volume for 150 mm × 25 mm petri dish).

5. Carefully, seed 3 ml of suspended cells directly on the slides.
6. Incubate cells under appropriate conditions to allow them to attach to the slides (*see* **Note 20**).
7. Wash the slides twice with ice-cold 1× DEPC-PBS and transfer the slides to a Coplin jar or slide transport box.
8. Fix the cells with freshly prepared 3.7 % formaldehyde (v/v) in 1× DEPC-PBS at room temperature for 20 min.
9. Discard the formaldehyde and wash the slides twice with 1× DEPC-PBS (*see* **Note 21**).
10. Dehydrate the cells by passing through an ethanol series (70, 85, and 99.5 % (v/v) in DEPC-ddH_2O, each step for 3 min).
11. Air-dry the slides and store at −80 °C (long-term storage) or −20 °C (up to 2 weeks) if so desired.
12. If slides have been stored, thaw at room temperature.
13. Attach Secure-Seal chamber(s) and rehydrate the cells by adding 1× DEPC-PBS-T to the chamber (*see* **Note 22**).
14. Remove DEPC-PBS-T and permeabilize the cells with 0.1 M HCl in DEPC-H_2O for 5 min.
15. Remove HCl and wash the cells twice with 1× DEPC-PBS-T (*see* **Note 23**).

3.2.2 Fresh Frozen Tissue

1. Tissue sections, mounted on microscope slides (*see* **Note 24**) are stored at −80 °C until use.
2. Thaw samples at RT.
3. Depending on specimen size, fix the tissue in the Secure-Seal chamber or a Coplin jar. Use 3.7 % formaldehyde in 1× DEPC-PBS for 45 min.
4. Wash once with 1× DEPC-PBS for 5 min.
5. (*Optional*) Permeabilize the tissue by incubating with pepsin (0.1 mg/ml in 0.1 M HCl at 37 °C for 5 min is a good starting point in our experience. Optimal conditions need to be identified for the respective specimen). *Preheat HCl to 37° C for optimal pepsin activity* (*see* **Note 25**).
6. Wash once with 1× DEPC-PBS for 5 min.
7. Dehydrate the tissue section in the ethanol series (70, 85, and 99.5 % ethanol in DEPC-ddH_2O, each for 3 min).
8. Air-dry and mount Secure-Seal chambers.
9. Rehydrate the tissue by adding 1× DEPC-PBS-T to the chamber.

3.2.3 Formalin-Fixed and Paraffin-Embedded (FFPE) Tissue

1. Tissue sections, mounted on microscope slides (*see* **Note 25**) are stored at −80 °C until use.

2. Thaw samples at RT.
3. Dewax samples by passing slides through a solvent series in Coplin jars:
   (a) Xylene for 15 min
   (b) Xylene for 10 min
   (c) Ethanol 100% for 2 min twice
   (d) Ethanol 95% for 2 min twice
   (e) Ethanol 70% for 2 min twice
   (f) Wash with DEPC-H2O for 5 min
   (g) Wash with DEPC-PBS for 5 min
4. Permeabilize the tissue by incubating with pepsin (0.1 mg/ml in 0.1 M HCl at 37 °C for 30 min is a good starting point in our experience. Optimal conditions need to be identified for the respective specimen.). *Preheat HCl to 37°C for optimal pepsin activity* (*see* **Note 25**).
5. Wash with DEPC-PBS for 5 min.
6. Postfix in 3.7% formaldehyde buffered in 1× DEPC-PBS for 10 min.
7. Wash twice with DEPC-PBS for 5 min.
8. Dehydrate the tissue section in the ethanol series (70, 85, and 99.5% ethanol in DEPC-ddH_2O, each for 3 min).
9. Air-dry and mount Secure-Seal chambers.
10. Rehydrate the tissue by adding 1× DEPC-PBS-T to the chamber.

3.3 mRNA Genotyping Protocol

The following protocol guides the user through the process of mRNA genotyping in cell lines, fresh frozen, formalin-fixed, and paraffin-embedded (FFPE) sections as well as tumor imprints after minor modifications [14]. We present the following protocol and volumes for a 50 μl reaction. Adjust volumes if necessary. Secure-Seal hybridization chambers are attached to the slides to isolate the specimen (*see* **Note 19**). At temperatures above room temperature, Secure-Seal chamber inlets should be covered with PCR film to prevent evaporation of reaction mix. Additionally, all incubations and reactions are performed in a humidified box. Finally, in accordance with good experimentation practice, replicates are recommended, since variation in handling slides and cell lines may influence the final result.

3.3.1 Reverse Transcription

1. Prepare reverse transcription mix according to Table 2, and apply the mix to the chamber.
2. Seal the Secure-Seal chamber inlets and incubate the slides at 37 °C. The optimal incubation time needs to be determined

Table 2
Reverse transcription reaction components

Reagent	Final concentration	Volume (μl)
TRANSCRIPTME—reverse transcriptase 200 U/μl	Variable[a]	Variable
RT—reverse transcriptase buffer (10×)	1×	5
RIBOPROTECT—RNase Inhibitor 40 U/μl	0.8 U/μl	1
BSA 20 μg/μl	0.2 μg/μl	0.5
dNTPs mix 10 mM	0.5 mM	2.5
LNA primer/random decamers 100 μM	1 μM/5 μM	0.5/2.5
DEPC-ddH_2O		Fill up to a total reaction volume
Total		50

[a]We typically use 5 U/μl for cell lines and 20 U/μl for tissue sections

empirically. We perform reverse transcription for 1 h when using LNA-modified target-specific primers. Random decamer primed reactions are typically incubated overnight.

3. Wash the slides once with 1× DEPC-PBS-T.

3.3.2 Postfixation

Postfixation is a crucial step that cross-links the newly synthesized cDNA strand to adjacent chemical groups of proteins. Always use freshly prepared fixative solution. We routinely use 3.7% formaldehyde in 1× DEPC-PBS, prepared from either 37% methanol-stabilized stock solution or made from paraformaldehyde powder. As for reverse transcription, the specific concentration and incubation time and temperature should be optimized for every specimen. At room temperature, we typically fix cell culture for 10 min and tissue sections for up to 45 min. Wash twice with 1× DEPC-PBS-T.

Storage point: At this point, the protocol can be paused and samples can be stored for a couple of days at 4 °C in 1× DEPC-PBS.

3.3.3 mRNA Degradation, Padlock Probe Hybridization, and Ligation

1. Prepare the reaction reagents according to Table 3 and apply the mix to the chamber.
2. Seal the inlets of the Secure-Seal chambers.
3. Incubate the slide in the humidity box at 37 °C for 30 min, then transfer the slide to 45 °C and incubate for 45 min (*see* **Note 26**).
4. Wash the slide twice with 1× DEPC-PBS-T.
   Storage point: At this point, the protocol can be paused, and samples can be stored for a couple of days at 4 °C in 1× DEPC-PBS.

Table 3
Ligation reaction components

Reagent	Final concentration	Volume (μl)
TTh ligase 200 U/μl	1 U/μl	1.25
TTh ligase buffer (10×)	1×	5
Padlock probe(s) 2 μM	0.1 μM	2.5
RNase H 5 U/μl	0.4 U/μl	4
BSA 20 μg/μl	0.2 μg/μl	0.5
KCl 1 M	0.05 mM	2.5
Formamide 100%	20%	10
DEPC-ddH_2O		24.25
Total		50

Table 4
Amplification reaction components

Reagent	Final concentration	Volume (μl)
Phi29 DNA polymerase 10 U/μl	1 U/μl	5
Phi29 DNA polymerase buffer (10×)	1×	5
dNTPs mix 10 mM	0.25 mM	1.25
BSA 20 μg/μl	0.2 μg/μl	0.5
Glycerol 50%	5%	5
DEPC-ddH_2O		33.25
Total		50

3.3.4 Rolling Circle Amplification

1. Prepare RCA mix according to Table 4 and apply the mix to the chamber.
2. Seal the inlets of the Secure-Seal chambers.
3. Incubate at 37 °C for ≥ 1 h (*see* **Note 27**).
4. Wash the slide twice with 1× DEPC-PBS-T.
   Storage point: At this point, the protocol can be paused and samples can be stored for a couple of days at 4 °C in 1× DEPC-PBS.

3.3.5 Decorator Probe Hybridization and Nuclei Counterstaining

IMPORTANT: *from this step onwards, fluorophore-conjugated probes and DNA intercalating dyes will be used. Protect decorator*

Table 5
RCP and nuclei staining reaction components

Reagent	Final concentration	Volume (μl)
Decorator probe(s) 10 μM	0.1 μM	0.5
2× Hybridization mix	1×	25
Hoechst 33342 100 mM	3 mM	1.5
DEPC-ddH_2O		23
Total		50

probes from direct light exposure for prolonged amount of time. Samples should be kept in a dark during and after incubation.

1 Prepare hybridization mix according to Table 5 and apply the mix to the chamber.

2 Incubate the slide at RT, protected from light, for ~20 min (*see* **Note 28**).

3 Wash the sample twice with 1× DEPC-PBS-T.

4 Mark the position of the chamber with diamond pen on the backside of the slide and remove the Secure-Seal chamber.

5 Dehydrate the specimen and remove glue residues and other contaminants from the slide by passing through an ethanol series (70, 85, and 99.5 % ethanol, each for 3 min).

6 Once the slide has dried, mount the coverslip with Slow-Fade medium. RCPs and cells are stable for a long time when kept at 4 °C and protected from light.

3.3.6 Image Acquisition and Analysis

Choose an appropriate imaging setup. We mostly use conventional wide-field epifluorescence microscopes to image tissue sections and cells. Depending on the level of detail required, select an appropriate objective (we typically use 20× and 40× high numerical aperture objectives). Avoid saturation when adjusting exposure times to allow for accurate signal segmentation during image analysis. Since the thickness of cells and tissue sections typically exceeds the depth of focus of the used objective we acquire Z-stacks with multiple focal planes that are combined to a single maximum intensity projection (MIP).

We routinely use the open-source cell image analysis software CellProfiler to quantify RCP signals [15] that can be accessed from the developer website (*see* **Note 29**), but other software packages can be used.

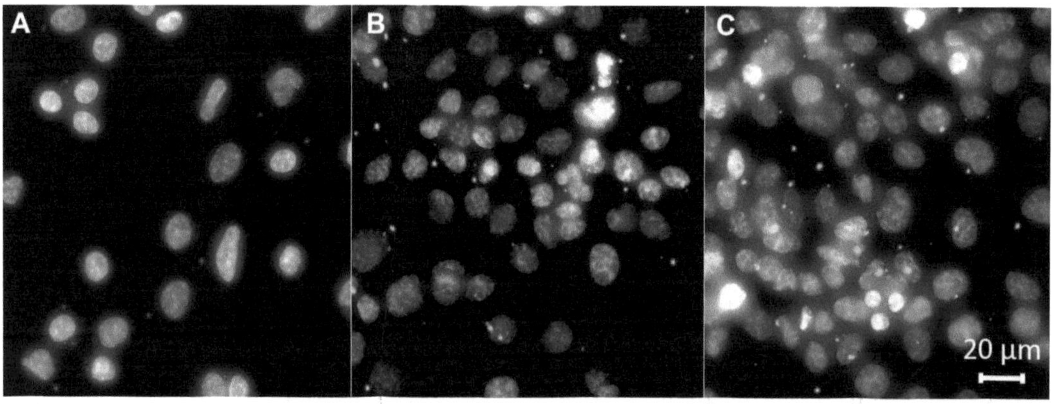

Fig. 3 In situ mutation detection of *KRAS c.34G>A* in cell lines. (**a**) RCPs originating from detection of a mutant KRAS allele in A549 cell line are represented as *red* speckles. Cell nuclei are shown in *gray*. (**b**) Detection of wild-type KRAS allele (green speckles) in the same position is presented in ONCO-DG1 cells. (**c**) A549 cells were spiked into ONCO-DG1 in 1:100 ratio. A mutated cell is depicted in the middle. Scale bar, 50 µm

3.4 Two examples of results

Each discrete fluorescent object represents a labeled RCP, originating from a hybridized and ligated padlock probe. During washes, cDNA molecules can diffuse out of cells and generate RCPs on the glass slide, thus extracellular RCPs are observed occasionally. Figure 3 shows *KRAS* mRNA genotyping in A549 and ONCO-DG1 cell lines. While the latter carries a wild-type *KRAS* allele, A549 cell line has a G>A mutation in position 34. When padlock probes for both alleles are used in parallel (*see* Table 1), either mutant or wild-type signals are present when the cell lines are stained individually. The method also allows for identification of single A549 cells spiked in ONCO-DG1 at a ratio of 1:100.

Examples where padlock probes were used to identify *KRAS* codon 12 and codon 13 mutations in FFPE tissue sections and tumor imprints are presented in Fig. 4.

4 Notes

1. To maximize cDNA synthesis efficiency, we recommend using target-specific primers where 5–7 nucleotides were changed with their LNA (locked nucleic acids; exact number of modified bases depends on the primer secondary structure) analogs. LNA bases should be interspaced with conventional DNA bases, beginning from the primer 5′. LNA bases not only display higher DNA hybridization affinity [16] but also protect target mRNA from RNaseH degradation, thereby fixing cDNA to the target mRNA.
2. *Tth* ligase is a well characterized enzyme that shows specificity and stability superior to other DNA ligases [12]. T4 DNA ligase can be used for conventional detection or mRNA, but mRNA genotyping can be compromised.

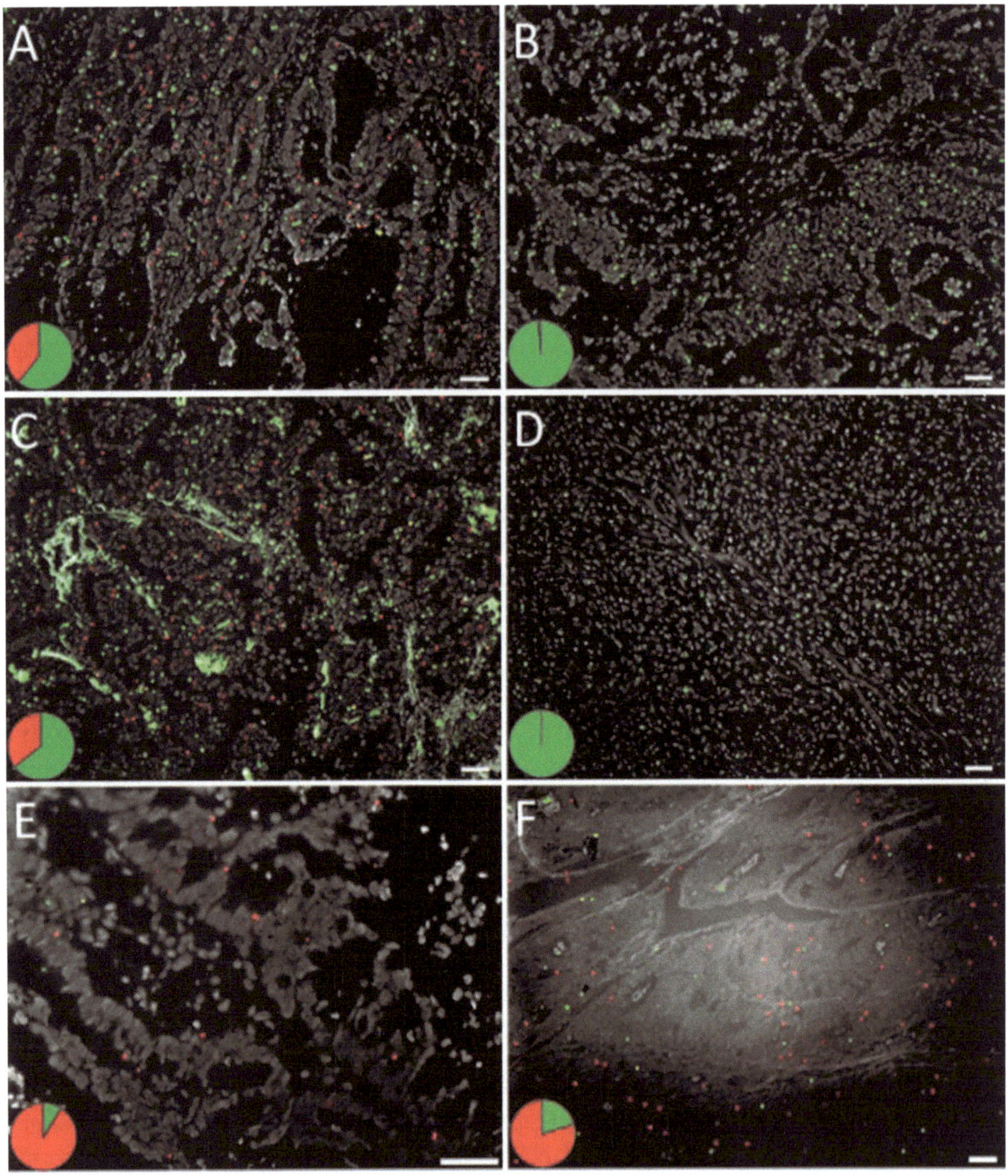

Fig. 4 In situ mutation detection of codon 12 and 13 *KRAS* mutations on (**a**–**d**) fresh frozen colon and lung tissues and (**e**, **f**) FFPE colon tissues using padlock probes and RCA. The tissues display *KRAS* mutant (*red*) and wild-type (*green*) RCPs. Cell nuclei are shown in *gray*. *KRAS* G12D mutation analysis in fresh frozen (**a**) mutant and (**b**) wild-type colon tumor tissue, in (**c**) mutant and (**d**) wild-type lung tumor tissue, and on FFPE colon tissues with reported (**e**) G12C or (**f**) G13D *KRAS* mutations. The *pie charts* indicate the ratio between wild-type (*green*) and mutant (*red*) signals in respective tissue. The images were acquired with 10× or 20× objective. Scale bar, 50 μm. Figure reproduced from [14]

3. In our experience, random decamers or dodecamers work best. Shorter oligonucleotides have a Tm below the temperature used during reverse transcription. Longer primers can form hairpins or dimers with other oligonucleotides in the pool. Standard desalting is a sufficient purification method. Reverse transcription of whole RNA content will create a cDNA pool that can hybridize with decorator probes, and lead to artifacts or elevated fluorescent background.
4. https://primer3.ut.ee [17] is a good starting point for PCR primer design. The webtool takes a sequence query from the user and suggests primers (only "right" primer should be considered) that meet sequence and structure criteria. Once primer sequence is known, we advise to check the primer for secondary structures (The OligoAnalyzer 3.1 tool from Integrated DNA Technologies (IDT), http://eu.idtdna.com/calc/analyzer, is a good secondary structure prediction engine). If there are any such structures predicted, LNA bases should not be introduced within structured regions. Finally, primers should be checked for nonspecific hybridization to prevent false-positive signals (nucleotide blast against Refseq mRNA database is a good choice, http://blast.ncbi.nlm.nih.gov/Blast.cgi). LNA-modified primers can be purchased from EXIQON (http://www.exiqon.com).
5. In our experience, increasing the distance between the SNV and the primer hybridization site typically results in a reduction in signal amount, putatively due to a decrease in target mRNA-to-cDNA conversion.
6. ProbeMaker allows for the automated design of allele-specific padlock probes. Additionally, linker/backbone elements and hybridization parameters can be specified.
7. Reaction constituents such as mono- and divalent ion concentrations, presence of formamide, probe concentration, and temperature influence the Tm of padlock probe arms. The hybridization and ligation reactions outlined in this chapter contain 75 mM monovalent ions, 10 mM divalent ions, 20% formamide, 0.1 μM padlock probe concentration, and is performed at 45 °C.
8. Shortened linkers presumably require target cDNA to bend to enable circularization of padlock probes, thereby potentially impairing hybridization and ligation.
9. Unique reporter motifs can be designed for each probe targeting different mRNA targets. Reporter motif and the corresponding decorator probe have the same, 20–25 bp long sequence. Decorator probe can be fluorophore-conjugated. We routinely use 6-Carboxyfluorescein, Texas Red, as well as multiple Cyanine and Alexa Fluor® dyes. Decorator probes

should be checked for nonspecific hybridization to minimize background fluorescence.

10. Secondary structure predictions of the padlock probe should be adjusted for the assay conditions (as with hybridization). The highest ΔG^0 is desired. We avoid loops and hairpin structures within padlock probe arms (as they can hinder hybridization of arms) or the reporter motif (as they can hinder hybridization of the decorator probe). Target sites should be blasted for off-target seqeunces to prevent false-positive signals. If multiple padlock probes are to be combined in an assay, their sequences are to be designed such that cross hybridization is avoided. The following protocol provides a guideline for probe phosphorylation: 10 μM final concentration of the padlock probe; 0.2 U/μl of T4 PNK kinase; 1× PNK kinase buffer A; 1 mM ATP and H_2O in a final volume of 50 μl. Mix should be incubated at 37 °C for 30 min, followed by enzyme inactivation at 65 °C for 20 min. Phosphorylated padlock probes can be stored in −20 °C until used.
11. We have used enzymes from NEB or Fermentas (Fisher Scientific), and they have performed equally well in our hands. This includes reverse transcriptase, RNase inhibitor, phi29 polymerase, and RnaseH.
12. All samples (sectioned tissues or fixed cells) should be stored at −80 °C to prevent RNA degradation.
13. DEPC is a nonspecific inhibitor of RNases present in water, buffers, or labware by irreversible covalent modification of selected amino acids [18]. DEPC is carcinogenic and should be handled with extra care (fume hood, gloves). Following DEPC treatment, solutions should be autoclaved to inactivate DEPC. Less dangerous chemical alternatives to DEPC, such as DMPC, can be considered.
14. Formamide is a known teratogen, irritating for skin and eyes. Handle with extra care (fume hood, nitrile gloves).
15. Hydrochloric acid is highly corrosive. Work under a fume hood with rubber PVC gloves.
16. Formaldehyde solutions in PBS should be freshly prepared from powder. We recommend aliquoting 3.7% formaldehyde in DEPC-PBS to 1 ml (used during the experiment) as well as larger volumes (15 ml) for cell fixation. Formaldehyde is a known carcinogen. Contact with skin, eyes, and clothes should be avoided. Use nitrile gloves and handle powder in the chemical fume hood.
17. Activity of lyophilized pepsin batches may vary, even from the same supplier. We recommend testing every batch for pepsin activity.
18. We use 0.1% v/v DEPC to treat PBS and ddH_2O for at least 2 h at 37 °C (or overnight at RT), followed by autoclaving.

19. Secure-Seal chambers of different sizes, shapes, and depths. For experiments performed on cells, we typically use 50 μl chambers (9 mm diameter, 0.8 mm deep). For larger areas or larger tissue specimen, 100 or 350 μl chambers can be used.
20. To maximize cell yield per slide, optimal conditions should be identified experimentally. In our experience, mostly based on work with immortalized human and mouse cell lines, overnight incubation allows cells to adhere to slides efficiently. Extended incubation can result in cell proliferation on-slide (clumped cells are difficult to segment by image analysis) while shorter incubation times can lead to premature termination of cell adherence.
21. Formaldehyde, larger quantities of concentrated HCl or formamide should be disposed in a safe manner, in accordance with local lab regulations.
22. Tween 20 as a surfactant will coat the chambers and ease swapping of buffers. Its presence in washing buffer can provoke bubble formation. Exchanging liquids in the chamber requires practice.
23. It is possible to suspend cells in 1× DEPC-PBS if one cannot proceed with the experiment immediately. In such case, cover the chamber inlets to prevent evaporation and keep the slide in 4 °C for up to 2 h.
24. We advise putting freshly cut tissue sections on slides that provide electrostatic attraction of cytological samples (SuperFrost Plus® from Menzel-Gläser work very well in our hands).
25. The fixation step should be optimized for every tissue type and thickness. The fixation needs to balance optimal reagent diffusion and minimize loss of tissue content. Take extra time to titrate fixation time (on consecutive sections), starting with a short incubation time. Use conditions showing maximal signal amount.
26. RNase H, that has the highest activity at 37 °C, will degrade mRNA from mRNA/cDNA heteroduplex within 30 min. The optimal temperature for Tth ligase is about 45 °C. Formamide, as a common nucleic acid destabilizer, lowers padlock probe Tm. This allows for using longer target recognition arms and in our hands, such an approach will increase the assay efficiency.
27. The recommended temperature for phi29 polymerase is 37 °C. Distinct RCPs can be observed after 1 h. If RCA is performed for several hours (overnight) at 37 °C, RCPs can start to fragment. If big RCPs are desired (dense tissues with high autofluorescence), we advise doing RCA at RT (overnight). Generally, optimal conditions for RCP generation have to be determined experimentally.
28. In a multiplexed reaction (when more than one detection oligo is used), we recommend hybridizing decorator probes at 37 °C for 30 min to minimize nonspecific binding of oligonucleotides.

29. CellProfiler is a powerful tool designed for biologists for image processing and analyses. It has multiple useful functions, including cell segmentation (e.g. to define nucleus and cytoplasm), speckle annotation or fluorescence measurement, and assigning the signal to the cell. A comprehensive manual and tutorials are provided on the developer website (http://www.cellprofiler.org). A pipeline to analyze the image set provided in Fig. 3 can be found in the "examples" tab, "Speckle counting". Analysis results can be exported into a *.csv* file for further processing.

Acknowledgements

We thank Evangelia Darai for conducting the A549/ONCO-DG1 genotyping experiment and providing images shown in Fig. 3.

References

1. Hanahan D (2014) Rethinking the war on cancer. Lancet 383:558–563
2. Hanahan D, Weinberg RA (2011) Hallmarks of cancer: the next generation. Cell 144:646–674
3. Femino AM, Fay FS, Fogarty K, Singer RH (1998) Visualization of single RNA transcripts in situ. Science 280:585–590
4. Lubeck E, Cai L (2012) Single-cell systems biology by super-resolution imaging and combinatorial labeling. Nat Methods 9:743–748
5. Player AN, Shen LP, Kenny D et al (2001) Single-copy gene detection using branched DNA (bDNA) in situ hybridization. J Histochem Cytochem 49:603–612
6. Choi HMT, Beck VA, Pierce NA (2014) Next-generation in situ hybridization chain reaction: Higher gain, lower cost, greater durability. ACS Nano 8:4284–4294
7. Nilsson M, Malmgren H, Samiotaki M et al (1994) Padlock probes: circularizing oligonucleotides for localized DNA detection. Science 265:2085–2088
8. Nilsson M, Banér J, Mendel-Hartvig M et al (2002) Making ends meet in genetic analysis using padlock probes. Hum Mutat 19:410–415
9. Fire A, Xu SQ (1995) Rolling replication of short DNA circles. Proc Natl Acad Sci U S A 92:4641–4645
10. Banér J, Nilsson M, Mendel-Hartvig M, Landegren U (1998) Signal amplification of padlock probes by rolling circle replication. Nucleic Acids Res 26:5073–5078
11. Stenberg J, Nilsson M, Landegren U (2005) ProbeMaker: an extensible framework for design of sets of oligonucleotide probes. BMC Bioinformatics 6:229
12. Luo J, Bergstrom DE, Barany F (1996) Improving the fidelity of Thermus thermophilus DNA ligase. Nucleic Acids Res 24:3071–3078
13. Zuker M (2003) Mfold web server for nucleic acid folding and hybridization prediction. Nucleic Acids Res 31:3406–3415
14. Grundberg I, Kiflemariam S, Mignardi M et al (2013) In situ mutation detection and visualization of intratumor heterogeneity for cancer research and diagnostics. Oncotarget 4:2407–2418
15. Carpenter AE, Jones TR, Lamprecht MR et al (2006) Cell Profiler: image analysis software for identifying and quantifying cell phenotypes. Genome Biol 7:R100
16. Petersen M, Wengel J (2003) LNA: a versatile tool for therapeutics and genomics. Trends Biotechnol 21:74–81
17. Untergasser A, Cutcutache I, Koressaar T et al (2012) Primer3-new capabilities and interfaces. Nucleic Acids Res 40:1–12
18. Wolf B, Lesnaw JA, Reichmann ME (1970) A mechanism of the irreversible inactivation of bovine pancreatic ribonuclease by diethylpyrocarbonate. A general reaction of diethylpyrocarbonate. A general reaction of diethylpyrocarbonate with proteins. Eur J Biochem 13:519–525

Chapter 5

The MassARRAY® System for Targeted SNP Genotyping

Justine A. Ellis and Benjamin Ong

Abstract

Research to understand the genetic basis of disease, particularly complex disease, regularly involves single nucleotide polymorphism (SNP) genotyping. The use of genome-wide SNP genotyping arrays has become increasingly more commonplace for gene discovery. However, smaller-scale genotyping platforms capable of efficiently genotyping tens to hundreds of SNPs are still crucial for many aspects of this work, including replication of associations. The Agena Bioscience MassARRAY System is one such platform. Here, we provide a guide to using the MassARRAY System, from assay design, through mass spectrometry, to generation of genotype data.

Key words Single nucleotide polymorphism (SNP), Genotyping, Mass spectrometry, MassARRAY, Polymerase chain reaction (PCR), Primer extension reaction, Multiplexing

1 Introduction

The shift from candidate gene to genome-wide approaches to genetic association studies has been swift and highly successful [1, 2]. The vast majority of new discoveries of genes associated with human complex diseases in the last decade have arisen from hypothesis-free genome-wide association studies (GWAS). GWAS approaches make use of single nucleotide polymorphism (SNP) arrays, where the selection of SNPs (anywhere from 500,000 to 5 million) is usually predetermined by the manufacturer. Despite the utility of these arrays for gene discovery, there is still an important place for platforms that allow genotyping of a bespoke selection of SNPs at targeted regions of the genome in a cost-effective manner. For example, following discovery of SNPs associated with a particular phenotype in a GWAS study, replication of findings will usually be required in a second population. Often only tens of SNPs require genotyping for this phase. Or perhaps greater variant coverage of a particular gene region is required in order to identify likely functional variants, or in-depth analysis of a select pathway of genes is required. In all of these examples, the vast majority of data generated by a GWAS array would be superfluous and an inefficient use of resources.

Stefan J. White and Stuart Cantsilieris (eds.), *Genotyping: Methods and Protocols*, Methods in Molecular Biology, vol. 1492,
DOI 10.1007/978-1-4939-6442-0_5, © Springer Science+Business Media New York 2017

The Agena Bioscience MassARRAY® System (formerly known as the Sequenom MassARRAY) is one such platform that allows for tens to hundreds of user-defined SNPs to be genotyped in hundreds to thousands of DNA samples in a high-throughput and cost-effective manner. Selected SNPs are assembled into groups of up to 40 that are compatible for the design of multiplex PCR assays. The assay design process is assisted by an online suite of programs that allow for design using 'default' settings, or more advanced manipulation by experienced users. Up to two plates of 384 samples can be genotyped for a 40-plex assay in around 10 h, resulting in the generation of 30,720 genotypes. Here, we provide an overview of the technology, and a step-by-step user guide to using the system for SNP genotyping based on many years' experience in our laboratory setting. Note that the MassARRAY System has other applications, such as measurement of DNA methylation, which will not be covered here.

1.1 Overview of Technology and Workflow

The MassARRAY is based on MALDI-TOF (matrix-assisted laser desorption/ionization—time of flight) mass spectrometry [3]. The multiplex (iPLEX®) assay procedure [4] is summarized in Fig. 1. Essentially, it employs PCR to amplify the regions of the genome containing each SNP. An 'extension' PCR reaction is then performed, in which an extension primer anneals just proximal to the polymorphic base, and a single 'terminator' nucleotide base extends the DNA fragment by one additional base that is specifically complementary to the polymorphic base. The terminator base, which lacks a 3′-hydroxyl group, prevents any further nucleotides from further extending the DNA fragment. The terminator bases are also 'mass-modified' so that mass differences between fragments differing by a single base are detectable by mass spectrometry. The expected mass for the fragment, dependent on which polymorphic base is present, can therefore be calculated.

The resultant multiplex analyte mixture is transferred to a SpectroCHIP® Array using a purpose-built dispenser such as the Agena Bioscience RS1000 Nanodispenser [5]. The SpectroCHIP® Array is pre-spotted with a matrix material to accommodate up to 384 individual analytes. The SpectroCHIP array is then placed in the MassARRAY mass spectrometer [6], and UV laser light is fired in short pulses at each SpectroCHIP spot (referred to as a 'pad') containing analyte/matrix co-precipitate, causing desorption and ionization (Fig. 2). A high-voltage electrostatic field forces the ionized DNA molecules to accelerate from the bottom of the vacuum tube to the top. Lighter ions travel faster and hit the detector at the top of the tube earlier than the heavier ions. After each laser pulse, the detector records the relative time of flight of each analyte, from which the mass of the DNA fragment can be calculated, and the nucleotide base present at the polymorphic site determined. The entire process of laser firing to signal detection takes only a few milliseconds, so up to 384 samples can be analyzed in less than 50 min.

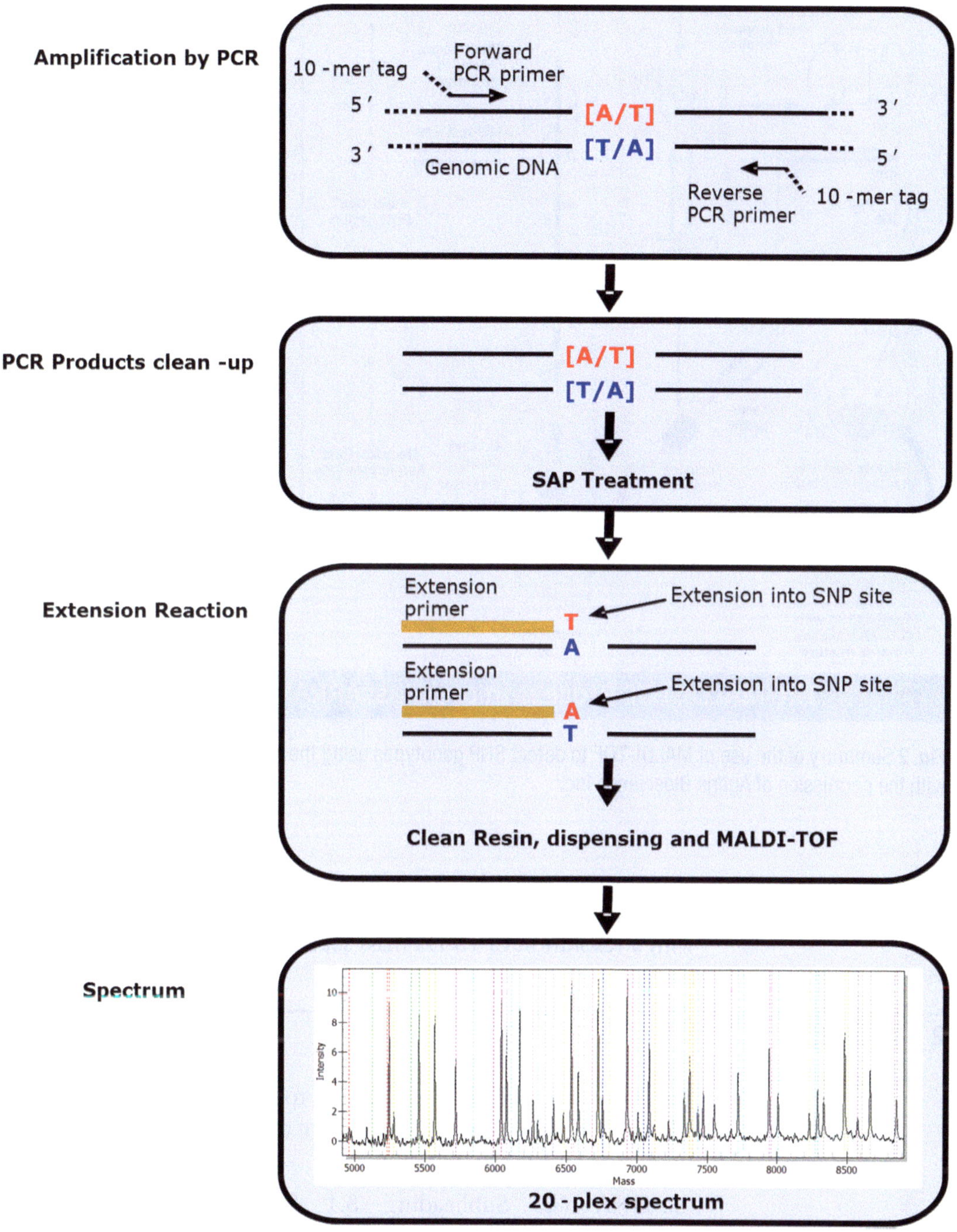

Fig. 1 Steps involved in the generation of SNP genotypes using the iPLEX chemistry. Regions targeted by the multiplex assay are amplified by PCR. PCR products are shrimp alkaline phosphatase (SAP) treated to neutralize unincorporated nucleotides. An extension reaction is then performed to extend the PCR fragments by one base into the SNP site. The mass of the resultant extended fragments are then measured using MALDI-TOF, resulting in a spectrum of distinct mass peaks for the multiplex reaction. Adapted from [4]

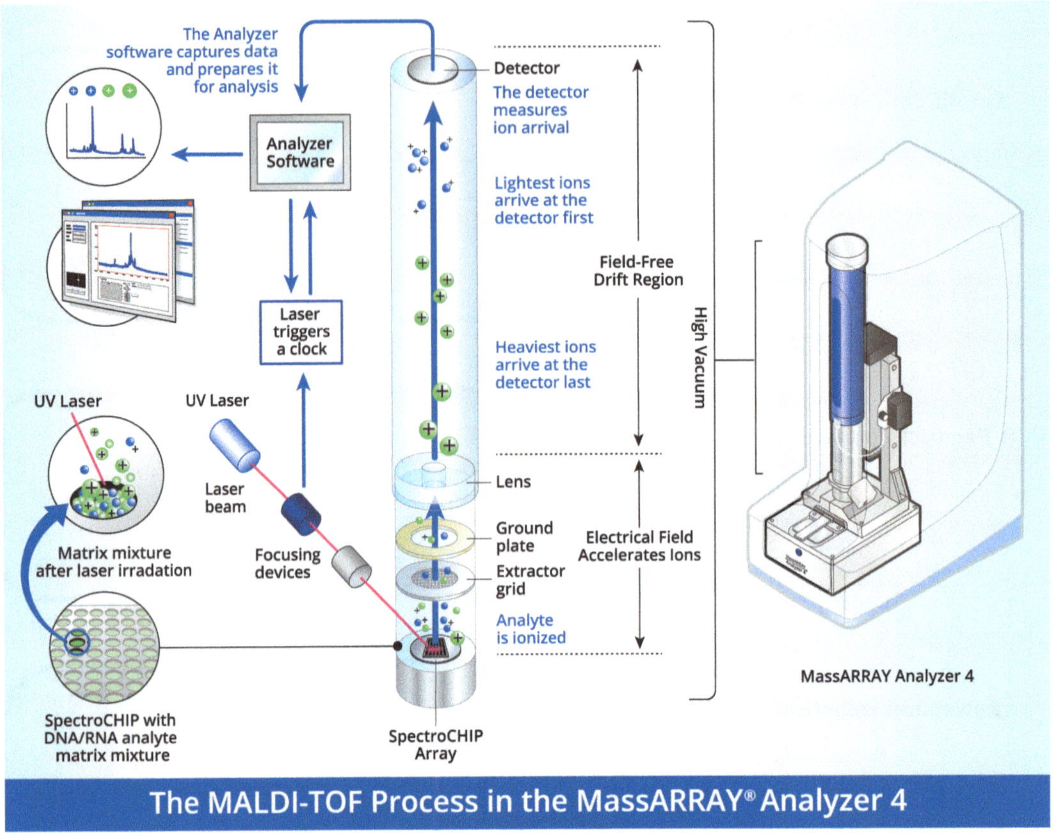

Fig. 2 Summary of the use of MALDI-TOF to detect SNP genotypes using the MassARRAY System. Figure used with the permission of Agena Bioscience Inc.

The MassARRAY Analyzer 4 System is designed to detect DNA within a mass range of approximately 4500 Da to 9000 Da, with a resolution of 16 Da mass separation [7].

2 Materials

The following materials, in addition to the MassARRAY MALDI-TOF and RS1000 Nanodispenser, are required (or recommended) to perform iPLEX genotyping:

1. DNA (*see* Subheading 3.1 for quality and quantity requirements).
2. PCR and extension primers (*see* Subheading 3.2 for design of appropriate primer sequences).
3. Complete iPLEX Gold genotyping reagent set (10 × 384) includes PCR reagents for amplification, shrimp alkaline phosphatase (SAP) enzyme and buffer, iPLEX Gold reagents for primer extension, SpectroCHIPs, and Clean Resin.

4. 384-well plates. Plates with the appropriate characteristics for use with the MassARRAY System can be obtained from Abgene. Only one 384-well plate is required to carry out the entire iPLEX process.
5. 96-well microtiter plates. If a robot is to be used for any pipetting to or from 96-well plates, these plates will need to be compatible with the robotic instrument.
6. Foil or film seals for microtiter plates.
7. Dimple plate—6 mg, 384-well, can be purchased from Agena Bioscience.
8. Ultrapure water such as Milli-Q water—deionized water with a resistivity of 18.2 MΩ cm.
9. Single, as well as 8 and/or 12 channel micropipettes (recommended).
10. Large volume repetitive dispenser pipette, such as the Eppendorf Repeater with Combitips (recommended).
11. 384-well thermal cycler capable of running the iPLEX thermal cycle program (*see* Subheading 3.6).
12. Centrifuge capable of spinning microtiter plates.
13. Microtiter plate rotator.
14. Centrifuge capable of spinning microtubes.

3 Methods

3.1 DNA Requirements for Successful Mass ARRAY Genotyping

DNA from a variety of biospecimen types has been successfully genotyped in our hands using the MassARRAY system. This includes DNA extracted from whole blood or white blood cell fractions, from cells collected by cheek brush, and from saliva (collected using specialized kits for DNA collection, or simply into sterile vials), chorionic villus samples and amniotic fluid. Suitable DNA can even be obtained from newborn blood spotted on to a card (Guthrie cards) and stored at room temperature for a number of years.

Superior genotyping success rates are generally achieved from the use of high-quality DNA (typically, a 260/280 ratio of 1.8 or greater) at a concentration within the system-recommended range of 5–10 ng/μl (*see* **Note 1**). To achieve this, particularly for more difficult biospecimen types such as Guthrie cards, it is worth trialing various DNA extraction methods to optimize outcomes. Methods designed to optimize yield and quality from a small amount of starting material, such as micro-column kits from companies such as Qiagen, can often be useful in this regard.

3.2 Assay Design

Here, we provide a brief overview of the basic steps to achieving an assay design using 'default' design settings (*see* **Note 2**).

1. Before attempting to design multiplex (iPLEX) assays for use on the MassARRAY System, a list of target SNPs should be assembled, identified by the reference SNP cluster ID, or 'rs' number. The selection of target SNPs is project-specific, and beyond the scope of this chapter.
2. Access Agena Bioscience's Assay Design software online [8]. New users must first register with Agena Bioscience, and to do so, users must be affiliated with an organization that is an Agena Bioscience customer. This should apply for all owners of the MassARRAY System.
3. Once registered and logged on to the site, launch the Assay Design Suite (version 2.0 as at October 2015) via the 'Online Tools' tab.
4. Start a 'new assay design' and input a project name. If the user has used the designer before, a list of previous assay designs will also be available.
5. Specify the SNPs to be included in the design (*see* **Note 3**). The selected SNP rs numbers should be assembled into a comma-separated list (using software such as Microsoft Excel) and uploaded using the 'file upload' button. Alternatively, if the list of SNPs is small, the rs numbers can be typed into the system directly using the 'edit text input' button.
6. Using the 'presets' dropdown menu, the level of iPLEX multiplexing needs to be specified. For lists of up to 12 SNPs, choose 'Low multiplexing iPLEX presets'. For lists of between 13 and 24 SNPs, choose 'Moderate multiplexing iPLEX presets'. For larger SNP lists, choose 'High multiplexing iPLEX presets'. Note that the 'multiplex level' box will display, by default, the maximum number of SNPs for the iPLEX level selected (but *see* **Note 4**).
7. Select the appropriate organism, and the genome database that should be used to extract information about the SNPs of interest, such as flanking sequence and other nearby sequence variants, necessary to carry out the design. For human research, the most recent human genome build available should be selected unless there is special reason to map SNPs to earlier builds.
8. Once all options have been specified, click the 'begin run' button (*see* **Note 5**). The program will automatically run through the various steps of the design process:
   (a) Retrieving and formatting SNP sequences.
   (b) Identifying proximal SNPs in the regions to be amplified that may interfere with primer binding.
   (c) Finding optimal primer locations.
   (d) Designing the assays by assembling compatible SNPs into groups suitable for multiplex reactions.

(e) Validating the assay by checking that all the various combinations of primers in the multiplex will not amplify unwanted regions of the genome.

9. Download the output files you will require for ordering primers and running your assay on the MassARRAY System. Click on the 'Design Assays' results button. A 'View Assay Design' window opens. The required files can be found by clicking the 'Export' button. *See* **Notes 6–8** for results file description/interpretation.

3.3 Ordering and Preparing PCR and Extension Primers

1. Order oligonucleotides according to the sequences contained in the 'oligo order file'. An amount of 25 nmole PCR primers, and 100 nmole extension primers should be ordered. All primers should be desalted to remove small molecule impurities, and delivered in lyophilized form.
2. The lyophilized PCR primers (forward and reverse) should be reconstituted to 100 μM, and the extension primers to 500 μM using Milli-Q water. These reconstituted primers are used as stock primers for downstream use, and should be stored in −20 °C freezers until required.

The extension primers will need to be pooled and adjusted before use in extension reactions (*see* **Note 9**).
There are two options for extension primer adjustment [9]:

1. A simple way to adjust the extension primers is to divide the primers into two groups of low and high mass. The concentration of the high mass group is then added to the extension reaction at double the concentration of those in the low mass group. For high plexes, the primers can be organized into three or four mass groups. Primers in the highest mass group are required to be at a concentration of 1.25 μM in the final iPLEX reaction mix.
2. While option 1 is quick and simple, it can result in less than optimal genotype call rates. Call rates can be improved by using an Excel spreadsheet ('Linear Primer Adjustment', available from Agena Bioscience) that utilizes a gradient algorithm to more accurately calculate the concentration of each individual oligonucleotide to equalize peak height.

3.4 Trailing New Extension Primers

The adjusted extension primer mixes should be tested on the MassARRAY System prior to use for genotyping. The Agena TYPER Analyser software [10] can then be used to create a primer adjustment report that recommends any further adjustment to the primer mixes to optimize downstream genotyping.

1. Make up a small volume (~100 μl) of the extension primer mix using either option 1 or 2 above.

2. Dilute the primer mix 1 in 10 using Milli-Q water.
3. Add 3×10 μl aliquots of each extension mix to a 384-well plate. Transfer the products onto SpectroChip and analyze on the MALDI-TOF. Please refer to Subheadings 3.7 and 3.8 for instructions.
4. TYPER Analyser creates a spectrum of peaks. The extension primers, which have not been incorporated into an extension reaction in this process, are referred to as 'unincorporated extension primers' or UEPs. The UEP peaks in the spectrum should be of even height, and no peaks other than UEPs should be present.
5. Access the Primer Adjustment Report via the File -> Reports menu. The Primer Adjustment report details recommended adjustments to the primer mixture for each assay in a multiplex. In a given well, the assay with the highest signal-to-noise ratio receives a score of 1 and the scores for other assays in the multiplex are computed relative to 1.

3.5 Quality Control

It is recommended to run a number of quality control reactions, alongside the genotyping reactions, to assess the potential for spurious peaks on the mass spectrum. These might include:

1. Inclusion of a DNA sample that is known to perform well for other iPLEX reactions. This assesses the performance of the iPLEX reaction in the presence of an 'optimal' DNA sample.
2. Inclusion of a well that has been subjected to both the PCR and iPLEX reaction steps, but does not contain any DNA (No Template Control, NTC). This assesses the likelihood of cross-contamination of DNA from other wells, along with the background spectrum generated by the presence of PCR and extension primers and other reagents alone.
3. Inclusion of a well in which Taq polymerase has not been added at the PCR stage. This assesses the background spectrum generated by the presence of DNA and primers subjected to both PCR and iPLEX protocols, but where no amplification of the target DNA sequence has occurred.

Each of these control wells should be run in duplicate (at a minimum) and placed in various locations across the 384-well plate, so that potential for background noise across the plate can be evaluated.

3.6 iPLEX Genotyping Process

3.6.1 PCR

Preparation of reagent mixes can be performed manually using single or (preferably) multichannel pipettes and repetitive dispenser pipette, or robotically using a liquid handling and dispensing robot. The protocol described below can be adapted for both manual and robotic pipetting.

The amount of Taq polymerase enzyme used depends on the plexing level. The amount of Taq used in a low plex assay, i.e. 26 SNPs or less, is half of that used in the high plex assay, i.e. 27 SNPs or greater.

1. Prepare the PCR primer mix. Reconstitute the forward and reverse primers at 100 μM in Milli-Q water and leave for several hours at room temperature, or overnight at 4 °C. Pool and dilute (with Milli-Q water) the PCR primers so that all primers are at a concentration of 0.5 μM within the volume required for the PCR reaction. For example, for a 30-plex, there are 60 primers. Divide the desired concentration (0.5 μM) by the reconstituted concentration (100 μM) and multiply by the volume required (e.g. 400 μl). Therefore, pool 2 μl of each reconstituted primer (total 120 μl primer) and bring total volume of pool to 400 μl by adding 280 μl Milli-Q water.
2. For 1 × 384 well plate, prepare the PCR master mix on ice as per Table 1. The Agena 'Complete PCR Reagent Set' can be used, or reagents can be sourced individually. It is recommended to make up sufficient master mix for 400 reactions to allow sufficient reagent 'overhang'.
3. Add 4 μl of PCR master mix to wells of an empty 384-well plate (hereinafter referred to as the 'analyte plate') using a repetitive dispensing pipette.
4. Add 1 μl of DNA at 5–10 ng/μl concentration to each reaction.
5. Centrifuge the analyte plate at 200 × *g* for 1 min.

Table 1

PCR master mix for a 384 well plate of low-plex and high-plex assays

Master Mix		Low Plex (≤26 SNPs)		High Plex (>26 SNPs)	
Reagent	**Conc. in 5 μl**	**×1**	**×400**	**×1**	**×400**
Milli-Q grade H_2O		1.9	760	1.8	720
Buffer	1×(2 mM $MgCl_2$)	0.5	200	0.5	200
$MgCl_2$[a]	2 mM	0.4	160	0.4	160
dNTPs	500 μM	0.1	40	0.1	40
Primer mix	100 nM	1.0	400	1.0	400
Taq Polymerase	0.5 U/1 U	0.1	40	0.2	80
Total		4 μl	1600 μl	4 μl	1600 μl

[a]Total $MgCl_2$ is 4 mM (2 mM from buffer, 2 mM from $MgCl_2$)

6. Place analyte plate in 384-well thermal cycler using the following program:

   94 °C for 4 min

   45 cycles of (94 °C 20 s, 56 °C 30 s, 72 °C 1 min)

   72 °C for 3 min

   4 °C hold
7. Proceed to remove unincorporated nucleotides.

3.6.2 Remove Unincorporated Nucleotides

Shrimp alkaline phosphatase (SAP) is used to neutralize unincorporated dNTPs in the PCR reaction. The SAP cleaves a phosphate from the unincorporated dNTPs, rendering them unsuitable for nucleotide addition in the iPLEX extension reaction.

1. Prepare the SAP mix on ice using the supplied reagents according to Table 2. It is recommended to make a mix sufficient for 410 wells, at 2 μl per well, for a 384-well plate.
2. Centrifuge analyte plate at 200 × *g* for 1 min and place plate on ice.
3. Add 2 μl of SAP master mix to each well of the analyte plate using a repetitive dispenser pipette.
4. Centrifuge analyte plate briefly to mix SAP mix with PCR products.
5. Place analyte plate in 384-well thermal cycler, and use the following program:

   37 °C for 40 min

   85 °C for 5 min

   4 °C hold
6. Proceed to iPLEX extension.

3.6.3 iPLEX Extension

1. For 1× 384-well iPLEX reactions, prepare the master mix on ice according to Table 3. It is recommended to make up sufficient master mix for 410 reactions (2 μl per well).

Table 2

Shrimp alkaline phosphatase (SAP) master mix for a 384-well plate

Master Mix	×1	×410
Water (MilliQ)	1.53	627.3
10 × buffer	0.17	69.7
SAP enzyme (1.7 U/μl)	0.3	123
Total	2 μl	820 μl

Table 3
iPLEX extension master mix for a 384-well plate of low-plex and high-plex assays

Master Mix	Low Plex (≤18 SNPs)		High Plex (>18 SNPs)	
Reagent	**×1**	**×410**	**×1**	**×410**
Water (MilliQ)	0.74	303.2	0.62	253.8
Buffer	0.2	82	0.2	82
Termination mix	0.1	41	0.2	82
Adjusted Primer mix[a]	0.94	385.4	0.94	385.4
iPLEX enzyme	0.02	8.4	0.04	16.8
Total	2 μl	820 μl	2 μl	820 μl

[a]Assumes extension primer mix has been prepared using the Linear Primer Adjustment method (spreadsheet available from Agena)

2. Centrifuge analyte plate briefly and place plate on ice.
3. Add 2 μl of iPLEX master mix to each well of the analyte plate using a repetitive dispenser pipette.
4. Centrifuge analyte plate briefly to bring reagents together.
5. Place analyte plate in 384-well thermal cycler, and use the following program:

   94 °C for 30 s

   40 cycles of (94 °C 5 s, (5 cycles of 52 °C 5 s, 80 °C 5 s))

   72 °C for 3 min

   4 °C hold
6. Proceed to de-salt the iPLEX products.

3.6.4 De-Salt the iPLEX Products

To remove salts from the iPLEX products prior to mass spectrometry, the Clean Resin ion exchange resin is used. This procedure requires a re-useable 'dimple plate'.

1. Add resin to the dimple plate wells using a spoon and scraper, ensuring all wells are full.
2. Allow to dry for around 20 min at room temperature (~25 °C). Do not over-dry.
3. While waiting for the resin to dry, add 16 μl of Milli-Q water to each well of the analyte plate.
4. Centrifuge the analyte plate at 300 × *g* for 1 min.
5. When the resin is dry, gently flip over the analyte plate so that it is upside-down on top of the dimple plate, with each well on

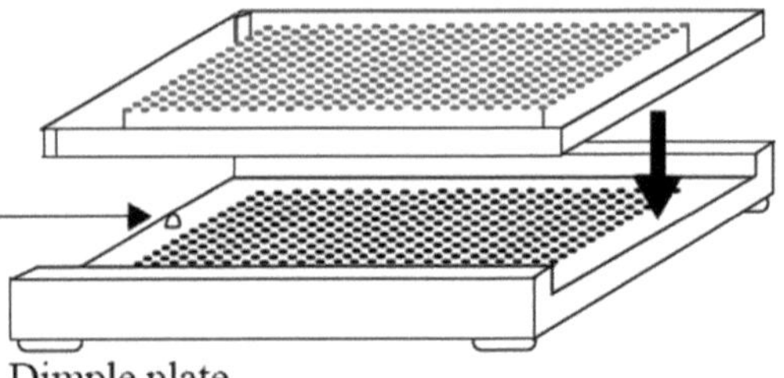

Fig. 3 Positioning of the 384-well analyte plate in the dimple plate for transfer of the Clean Resin ion exchange resin. Figure used with the permission of Agena Bioscience Inc.

the analyte plate aligned perfectly with each well of the dimple plate (Fig. 3). The solution/liquid in the analyte plate will not fall out as it adheres to the wall of the wells.

6. Holding the analyte plate and the dimple plate together, flip them over so that the dimple plate is now on top of the analyte plate. Gently tap the dimple plate to ensure all the resin falls into each well of the analyte plate. Seal the plate and briefly (pulse) centrifuge.
7. Rotate the analyte plate for a minimum of 5 min (up to 2 h) at room temperature.
8. Centrifuge the analyte plate at 3200 ×*g* for 5 min. The analytes are now ready for spotting to SpectroCHIP, but can be sealed and stored at −20 °C for up to 2 weeks before use.

3.7 Transfer of Analytes to SpectroCHIPs

Generally in a core facility, the transfer of analytes to SpectroCHIPs, and firing of analytes through the mass Spectrometer, are performed by trained personnel. As such, we provide only a brief overview of these procedures here.

The Nanodispenser RS1000 [5] or other compatible dispenser instrument is used to transfer resin-cleaned iPlex products (analytes) from 384-well plates to SpectroCHIPs. A maximum of 2 × 384-well analyte plates can be transferred at a time using the Nanodispenser.

1. Following centrifugation, place the analyte plate with well A1 to the lower left of one of two plate holders on the Nanodispenser. Place the SpectroCHIP on the chip position.
2. Load 30 μl of three-point calibrant (*see* **Note 10**) into the calibrant reservoir. Select the parameters for transferring the analytes:
   (a) Mapping: Select '384 plate to 384 chip' if transferring analytes from 384-well plate to 384-well formatted SpectroCHIP.

(b) Volume: The recommended mean volume of the droplets is 8–10 nl (SD ± 5 nl). To achieve acceptable mean volume, adjust the dispense speed to deposit smaller or larger droplets as required.

(c) Number of SpectroCHIPs; maximum 2 chips.

(d) Analyte and/or calibrant dispensing: both are required in a normal run.

(e) Dispensing speed (mm/s): determines the amount of analyte to dispense. The higher the speed the more analyte dispensed.

(f) Calibrant speed: set at 130–140 mm/s

(g) Cleaning: all options should be selected

3. Select Transfer and click Run. This will start the process of picking up the analyte from the 384-well plate and dispensing it onto the SpectroCHIP.

3.8 Acquiring Genotype Spectra Using the MassARRAY MALDI-TOF Analyser

Once nanodispensing is complete, set up the mass spectrometer run as follows:

1. Within the TYPER program suite, open the 'Assay Editor' program and upload the Assay Group file generated by the Assay Design Suite (*see* Subheading 3.2).
2. Define how assays and plates are set up in the MassARRAY database using the 'Plate Editor' program in TYPER. You will be required to enter DNA sample ID numbers and assay information for each well (drawn from the uploaded Assay Group file).
3. Connect this information to the MassARRAY mass spectrometer using the TYPER Chip Linker software.
4. Place the SpectroCHIP onto the MassARRAY analyser. Two SpectroCHIPs can be analyzed at a time.
5. In the MassARRAY real-time software (SpectroAcquire), load iPLEX parameter so that the MALDI-TOF is set up to run iPLEX genotyping samples.
6. Type in the barcode of the SpectroCHIP into the chip field. Click Barcode Report to confirm that the input barcode matches the created file.
7. On the Auto Run tab, click Run.

After completion, spectrum data are outputted as an .xml file, which can be viewed and analyzed using the TYPER Analyser software within the TYPER suite of programs [10].

3.9 Data Analysis

The TYPER Analyser program has many functions and contains many user-specified options. A full instruction in the use of TYPER Analyser is beyond the scope of this chapter. We recommend that the analysis of MassARRAY data be carried out with reference to the TYPER software User's Guide [10], which can be directly accessed from within TYPER Analyser under the Help menu. Here, we provide basic instructions to derive genotype information from spectrums generated by the MassARRAY.

The .xml file produced by the MassARRAY can be viewed and genotype information extracted as follows:

1. Open the TYPER software program.
2. Select 'TYPER Analyser'.
3. On the 'View' tab, ensure the following panes are visible within the TYPER window:
   (a) Project explorer
   (b) Traffic light
   (c) Chip summary
   (d) Call cluster plots
   (e) Post-processing clusters
   (f) Details
4. In the File menu, select 'open wells from file'.
5. Select the xml file to be analyzed. If not connected to the MassARRAY database, the file will need to be retrieved and saved to the local computer.
6. Select the chip name associated with the data to be analyzed. A 'traffic light' display of the 384-well analyte plate appears, along with a list of SNPs for the selected well.
7. A scatterplot of the datapoints for each SNP can be viewed by selecting the SNP to be viewed in the Assay pane, then clicking on the 'Call Cluster Plot' tab. Genotypes assigned to these datapoints by the basic 'Caller' software are also displayed. The Caller software relies primarily on the ratio of peak heights for the alleles, but does not take into account any other characteristics of the spectra that are specific to a particular assay. These can be better taken into account via cluster analysis.
8. To apply a cluster analysis (Gaussian mixtures approach) to the genotype calls for all SNPs, Select 'Autocluster' in the Tools tab.
9. Once clustering is completed, select the SNP of interest in the Assay pane then click on the 'post-processing clusters' tab. You should now see the final post-processing genotype clustering, along with a list of sample IDs and their genotype for that SNP. There are two columns containing genotype calls—the 'call' column, which contains the genotype called by the Caller software, and the 'cluster call' column which contains the genotype called by the clustering method.
10. Click on any datapoint in the cluster plot (*see* **Note 11**) to see the sample ID and genotype call associated with it. If you wish

to change the genotype call for that data point, right click on it, and select 'change call'. This is useful if you do not agree with the genotype assigned to that datapoint (e.g. if the datapoint sits clearly out of a cluster, or the peak intensity is very low) and you wish to discard the genotype call for that sample (select 'no call').

11. Click on the 'Details' tab to view the spectrum by sample ID. The position on the spectrum for the peaks of the two alternate alleles is marked, along with the location of the peak expected for unincorporated extension primer (UEP). A high-intensity peak for a UEP might indicate that the iPLEX extension reaction for that SNP has not been optimal.
12. Once satisfied with the genotype calls for each SNP, the genotype data organized by sample ID and SNP number can be saved by opening the 'Plate Data' pane (View menu) and clicking the 'save as' icon. The saved genotype data file will open in Microsoft Excel ready for downstream analysis (*see* **Note 12**).

4 Notes

1. The MassARRAY iPLEX system is relatively forgiving in terms of required DNA quality and quantity. In our experience, it is often worth attempting genotyping with poorer quality/quantity DNA samples if that is all that is available. It is generally true that poorer DNA samples may lead to less reliable genotype calls. However, simple quality control (QC) approaches to cleaning the genotyping data, such as removal of data from any DNA samples not achieving a genotyping rate of at least 90%, can easily be applied before the data are used in downstream statistical association analyses.
2. Refer to the current Assay Designer Software User Guide [11] for guidance on assay design. The appearance of the Assay Designer user interface may change from time to time, and thus it is always recommended to refer to the software user guide current at the time of use.
3. Around 5–10% of SNPs across the genome will fail the assay design process. In this case, it is often possible to identify and add a 'proxy' SNP to the assay, which is highly correlated (for example, a linkage disequilibrium r^2 of 0.8 or greater) with the failed target SNP. Searching for a proxy SNP can be simply done using web-based databases such as the Broad Institute's SNAP [12].
4. During assay design, the 'high multiplex' setting can be extended from the default of 36 SNPs, to 40 SNPs, without hampering the design process. The program will not accept an input greater than 40 in the 'multiplex level' box.

5. Each of the assay design steps can be run individually if preferred—this can provide opportunity to, e.g. assess rejected SNPs and make changes to the SNP list without having to wait until the entire design process has completed.
6. The Assay Designer 'oligo order' file provides a spreadsheet of primer sequences in a format directly accepted by some oligonucleotide manufacturers, streamlining the process of ordering large numbers of primers. The 'Assay group' file contains details of the number of multiplex assays that have been designed (termed 'wells' and denoted by W1, W2, W3, etc.) and detailed information about the PCR and extension primers, the amplicons they are expected to produce, and the mass of the extension products for each possible allele. The 'Design Summary' file provides a detailed technical overview of the settings used in the design, and the composition of each well. The 'failed sequences' file contains a list of SNPs that failed design and reports the reason for the failure (*see* **Note** 7). The 'SNP Group' file contains the flanking sequences of the SNPs that passed design and appear in one of the wells. Location of the SNP and of the primer sites is indicated in capital letters. The 'Assay Design Step Log' provides a log of the entire assay design process. It is good practice to save all file types, as some or all will be needed to run your assay or interpret your data.
7. SNPs may fail assay design for a variety of reasons. These may include an inability to identify primer positions (PCR or extension) of sufficient specificity to prevent amplification of non-target sequences, primer dimer or hairpin formation, or the presence of other SNPs in the target region. If the target region contains SNPs other than the SNP to be genotyped, the designer will attempt to identify primer regions that avoid inclusion of these SNPs in the primer sequences. This is because the presence of an alternate nucleotide base in the sequence can interfere with primer annealing, resulting in a bias towards successful genotyping of only those strands with the common nucleotide base. Many of the above issues can be resolved by altering the default design settings; for example, longer amplicons may be permitted, providing more options for PCR primer positioning, or other SNPs in the region with very low frequencies in the population can be ignored. For beginners, we recommend consulting with the helpful scientists at Agena Bioscience, or working with your MassARRAY facility manager to alter such settings without compromising the assay design.
8. There are methods available to consolidate 'wells', and hence reduce the number of multiplex assays. Consult Agena Bioscience for assistance.
9. The adjustment of extension primers is necessary because of the inverse relationship between peak intensity and analyte mass. The peak intensity of the highest mass (~8500 Da) is

25% less than the average of the lower mass primers [9]. Without adjustment this issue will pose (1) a significant challenge to the genotype caller software because of the signal-to-noise ratios, (2) analyte peaks can be missed, leading to genotyping errors, and (3) nonpredictable variations in peak heights can occur. These variations may stem from inconsistent oligonucleotide quality and poor desorption/ionization behavior in MALDI [9].

10. The three-point calibrant is used by the MALDI-TOF to establish the equation for the best-fit curve for sample data using three unique oligonucleotides of known mass (5045, 8480, and 9980 Da).
11. To scrutinize genotype calls within TYPER Analyser for low-intensity peaks, we suggest viewing the post-processing 'polar plot' rather than the 'cartesian' plot. To increase the stringency, and thus the accuracy of your genotyping calls, we recommend setting the clustering magnitude cutoff to '5' (higher than default). This will ensure a 'no call' result for low-intensity SNPs, which may be more prone to error. We also suggest manually checking all calls changed by the clustering analysis (i.e. considering the position of these datapoints in the clusters, and viewing the peak heights using the 'details' pane), and in particular all genotype calls marked 'moderate' or 'aggressive'. We recommend erring on the side of caution—if a datapoint is well outside a cluster, is of low intensity, or if the allele peaks are clearly not consistent with the genotype assigned, it is usually a good idea to 'fail' the assay for that DNA sample (change the genotype call to 'no call'), and re-genotype the sample in an additional MassARRAY run. Some samples will consistently fail, while others will genotype well on a second attempt.
12. While the statistical analysis (e.g. genetic association analysis) of the genotyping data is outside the scope of this chapter, we suggest that any SNP that has not achieved a genotyping call rate of at least 90% across all of the samples, be discarded from analysis. Similarly, after discarding failed SNPs, any DNA sample that has not achieved a genotyping call rate of at least 90% should also be discarded from analysis. This helps to ensure that only high-quality genotype data are used in downstream statistical analyses.

Acknowledgement

We thank the various laboratory personnel who have assisted with the development of our in-house protocols over time, particularly Dr. Anna Duncan and Mr. Raul Chavez. JAE is supported by an Australian Research Council Future Fellowship.

References

1. Kruglyak L (2008) The road to genome-wide association studies. Nat Rev Genet 9:314–318
2. Visscher PM, Brown MA, McCarthy MI et al (2012) Five years of GWAS discovery. Am J Hum Genet 90:7–24
3. Oeth P, del Mistro G, Marnellos G et al (2009) Qualitative and quantitative genotyping using single base primer extension coupled with matrix-assisted laser desorption/ionization time-of-flight mass spectrometry (MassARRAY). Methods Mol Biol 578: 307–343
4. Oeth P, Beaulieu M, Park C, et al. (2007) iPLEX Assay: Increased plexing efficiency and flexibility for MassARRAY system through single base primer extension with mass-modified terminators. Sequenom Application Note Document No. 8876–006
5. Agena Bioscience MassARRAY Nanodispenser RS1000 User's Guide. www.agenacx.com
6. Agena Bioscience MassARRAY Analyser Compact User's Guide. www.agenacx.com
7. Agena Bioscience iPLEX Chemistry Application Note. www.agenabio.com
8. AgenaCX. www.agenacx.com/Home
9. Agena Bioscience iPLEX Gold Application Guide. www.agenacx.com
10. Agena Bioscience Typer V4 User's Guide. www.agenacx.com
11. Agena Bioscience Assay Design Suite User's Guide. www.agenacx.com
12. Broad Institute SNAP. www.broadinstitute.org/mpg/snap/

Chapter 6

Targeted Capture and High-Throughput Sequencing Using Molecular Inversion Probes (MIPs)

Stuart Cantsilieris, Holly A. Stessman, Jay Shendure, and Evan E. Eichler

Abstract

Molecular inversion probes (MIPs) in combination with massively parallel DNA sequencing represent a versatile, yet economical tool for targeted sequencing of genomic DNA. Several thousand genomic targets can be selectively captured using long oligonucleotides containing unique targeting arms and universal linkers. The ability to append sequencing adaptors and sample-specific barcodes allows large-scale pooling and subsequent high-throughput sequencing at relatively low cost per sample. Here, we describe a "wet bench" protocol detailing the capture and subsequent sequencing of >2000 genomic targets from 192 samples, representative of a single lane on the Illumina HiSeq 2000 platform.

Key words Molecular inversion probes, Massively parallel sequencing, Real-time PCR, Exonuclease cleanup and gel electrophoresis

1 Introduction

The ability to selectively enrich thousands of genomic DNA targets and sequence them in parallel has tremendously impacted the way genomes can be interrogated on a large scale [1]. Molecular inversion probes (MIPs) represent one such approach based on target circularization of single-stranded oligonucleotides consisting of a common DNA backbone flanked by target-specific sequences [2] (Fig. 1). Following hybridization of site-specific targeting arms, non-strand displacing DNA polymerase and deoxynucleotides facilitate extension (gap-closure) between targeting arms and the intervening sequence. The addition of DNA ligase completes the covalently closed circular molecule and exonuclease treatment removes linear DNA that failed to form a closed circle. PCR using universal primers complementary to the MIP backbone completes the DNA capture reaction and the library is, in principle, ready for DNA sequencing [3, 4].

These authors contributed equally to this work.

Stefan J. White and Stuart Cantsilieris (eds.), *Genotyping: Methods and Protocols*, Methods in Molecular Biology, vol. 1492, DOI 10.1007/978-1-4939-6442-0_6, © Springer Science+Business Media New York 2017

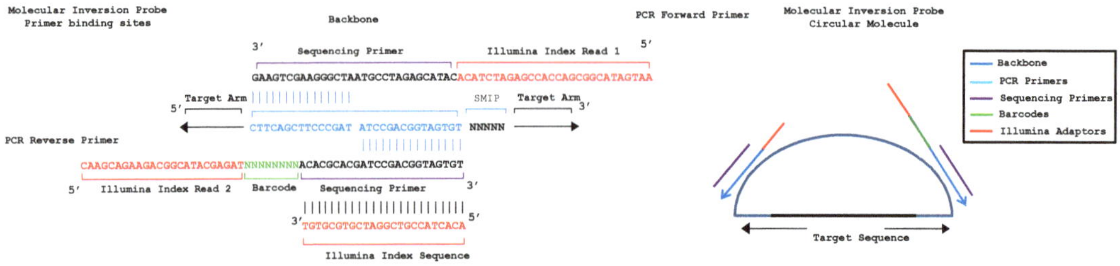

Fig. 1 Text-based primer map for molecular inversion probes. Sequence overlaps are annotated against the MIP backbone (*blue*). Positions of the forward and reverse PCR primers are annotated in *black* with Illumina index primers annotated in *red*. Forward and reverse sequencing primers (*purple*) overlap the MIP backbone and the MIP PCR primers. The 8 bp sample-specific barcode is annotated in *(green)* and the small molecular tag in *(grey)*

The success of any targeted enrichment approach is directly impacted by the performance of the DNA capture reaction. The MIP protocol has proven to be adaptable and the integration of recent technical advances has led to notable improvements in MIP performance [4–7]. MIPs demonstrate consistent capture uniformity (~98 % of captured targets), capture specificity (>99 % target overlap), and multiplex scalability (thousands of capture targets) [4, 5]. Improvements to MIP-design tools also allow *in silico* predictions of assay success leading to increases in capture efficiency [6]. In addition, the use of single-molecule tagging, by adding random unique barcode tags to each molecule (termed smMIPs), has also facilitated the quantitation of individual capture events, allowing for highly sensitive variant calling and precise quantitation of somatic or mosaic events [7]. The simplicity of the workflow procedure, low sample input requirements, and cost-effectiveness of the MIP protocol have proven advantageous for the detection of rare and *de-novo* mutations in large disease cohorts [5, 8]. This protocol therefore describes in detail a method for large-scale resequencing of several thousand genomic targets using MIPs.

2 Materials

Prepare all dilutions using nuclease-free water. All enzymes and mastermixes should be stored at −20 °C unless otherwise noted. All reagents should be thawed and prepared on ice unless otherwise noted. All waste disposal regulations must be followed when disposing of hazardous materials.

1. Components for MIP pooling and phosphorylation.
   (a) 70mer oligonucleotides synthesized at the 25 nanomole (nM) scale and hydrated to 100 micromole (μM) in 1× TE Buffer, pH 8.0. Store at −20 °C.
   (b) T4 DNA Ligase Reaction Buffer with 10 mM ATP (New England Biolabs). Store at −20 °C.

   (c) T4 Polynucleotide Kinase. Store at –20 °C.
   (d) ABgene 8-Flat-Cap Strip Tubes.
   (e) Costar* Microcentrifuge Tubes 1.7 mL; Color: Natural (holds 1.5 mL).
   (f) Nuclease-free water or equivalent. Store at room temperature.
2. Components for targeted capture.
   (a) Ampligase 10× Reaction Buffer (Epicentre). Store at –20 °C.
   (b) Ampligase® DNA Ligase (Epicentre). Store at –20 °C.
   (c) Hemo Klentaq (New England Biolabs). Store at –20 °C.
   (d) 10 mM Deoxynucleotide (dNTP) set. Product should be diluted fresh for each capture reaction 1:40 (0.25 mM). Store at –20 °C.
   (e) Nuclease-free water or equivalent. Store at room temperature.
   (f) Phosphorylated MIP pool (*see* **Notes 1** and **2**).
   (g) Eppendorf skirted 96-well plates or equivalent (clear).
   (h) Thermo Scientific* ABgene* Adhesive PCR Film or equivalent.
3. Components for exonuclease treatment.
   (a) Exonuclease I (*E. coli*). Store at –20 °C.
   (b) Exonuclease III (*E. coli*). Store at –20 °C.
4. Components for PCR.
   (a) iProof High Fidelity Master Mix (Bio-Rad). Store at –20 °C.
   (b) SYBR® Green I nucleic acid gel stain (Invitrogen). Store at –20 °C. Keep away from light.
   (c) Oligonucleotides. Synthesized at the 25 nM scale and hydrated to 100 μM in 1× TE Buffer, pH 8.0. Store at –20 °C.
   (d) Low-Profile 0.2 mL 8-Tube Strips Without Caps.
   (e) Optical Flat 8-Cap Strips.
5. Clean-up protocol components.
   (a) Agencourt AMPure XP beads. Store at 4 °C.
   (b) Ethanol (100%). Store at room temperature.
   (c) DynaMag™-2 magnet (ThermoFisher Scientific).
   (d) Buffer EB (Qiagen). Store at room temperature.
6. Agarose gel electrophoresis components.
   (a) E-Gel® EX Gel, 2% (Invitrogen).
   (b) E-Gel® Low Range Quantitative DNA Ladder (Invitrogen). Store at 4 °C.

7. Sequencing components.
   (a) Qubit dsDNA High Sensitivity Assay Kit. Store at room temperature.
   (b) 0.5 mL tubes (for Qubit) or equivalent.
   (c) Illumina MiSeq Reagent Kit (300 cycles PE). Store at −20 °C.

3 Methods

3.1 Oligonucleotide Pooling and Phosphorylation

1. Design MIPs using an existing pipeline [6] (*see* **Notes 1** and **2**).
2. Pool oligonucleotides at equimolar concentrations by plate, by combining 5 μL of each MIP (100 μM/μL) into a single 1.5 mL tube. Each individual 1.5 mL tube will represent a combined sum of 96 MIPs for a total volume of 480 μL (*see* **Note 3**).
3. Take 9.6 μL of each individual MIP pool (0.1 μL multiplied by the number of MIPs in each plate) and combine these into a single tube to generate a MIP megapool.
4. Phosphorylate the MIP megapool by combining 25 μL of the MIP megapool, 3 μL of 10× T4 DNA Ligase Reaction Buffer, 1 μL of T4 Polynucleotide Kinase (10 U), and 1 μL of nuclease-free water in a total reaction volume of 30 μL. Using a thermocycler, incubate the reaction mix at 37 °C for 45 min with a final denaturation step of 65 °C for 20 min. Store unphosphorylated MIPs at −20 °C for future use.

3.2 Targeted MIP Capture

1. Calculate the volume of the MIP megapool required in the capture reaction based on the ratio of desired MIP copies to DNA copies. This example will assume a megapool of 2000 MIPs captured using 100 ng of total genomic DNA, for a total ratio of 800 MIP copies to 1 DNA copy.
2. Calculate the expected number of MIP copies required given an input of 100 ng of genomic DNA, e.g., 800 × 33,000 haploid genome copies = 2.64×10^7 MIP copies required.
3. Transform the number of MIP copies to picomoles (pmol) using Avogadro's number (6.02×10^{23}), e.g., ($2.64 \times 10^7 / 6.02 \times 10^{23}$) ($1 \times 10^{12}$) = 4.38×10^{-5} pmol.
   Calculate the picomole per μL concentration of the MIP megapool:
   e.g., 0.1 μL × 100 μM/2000 MIPs = 0.005 μM
   (0.005 × 25 μL)/30 μL = 0.004 pmol/μL
4. Calculate the volume of 1× MIP megapool required in the capture reaction (*see* **Note 4**):
   e.g., 4.38 × 10−5 pmol/0.004 pmol/μL = 0.011 μL per capture reaction.
5. Prepare a 15 μL capture reaction on ice by combining 2.5 μL of Ampligase 10× Reaction Buffer, 0.0032 μL of 0.006 mM

dNTP mix, 0.32 μL Klentaq (10 U/μL), 0.01 μL of Ampligase (100 U/μL), 0.0105 μL of MIP megapool, and 12.16 μL of nuclease-free water. The total volume of DH_2O can be scaled depending on your DNA concentration requirements and the volumes are based on processing 192 samples (see below).

6. Plate 10 μL of DNA into a 96-well plate format (10 μL at 10 ng/μL). A range of 100–200 ng total DNA can be used in the final capture reaction.
7. Add 15 μL of capture reaction to each individual DNA sample.
8. Seal with adhesive PCR film (*see* **Note 5**).
9. Using a thermocycler, incubate the reaction mix at 95 °C for 10 min and 60 °C for 22 h. Remove plates and immediately place on chilling blocks (*see* **Note 6**).
10. Exonuclease treatment.
    (a) Immediately following capture, prepare a Exonuclease clean-up master mix containing 0.5 μL of Exonuclease I, 0.5 μL of Exonuclease III, 0.2 μL Ampligase 10× Reaction Buffer and 0.8 μL nuclease-free water per sample.
    (b) Add 2.0 μL of Exonuclease clean-up mix to each 25 μL capture reaction (*see* **Note** 7).
    (c) Using a thermocycler, incubate the reaction at 37 °C for 45 min and 95 °C for 2 min. Cool reaction plates to 4 °C (*see* **Note 5**).
    (d) Samples may be stored at 4 °C for a short term until PCR, or −20 °C for longer periods.
11. Real-Time PCR.
    (a) Prepare a RT-PCR master mix by combining 12.5 μL of 2×iProof High Fidelity Master Mix, 0.125 μL of 100 μM universal MIP barcode forward primer, 0.125 μL of 100× SYBR® Green I nucleic acid gel stain and 6.125 μL of nuclease-free water.
    (b) Add 18.75 μL of RT-PCR master mix to each well.
    (c) Add 1.25 μL of 10 μM individual barcode primers and 5 μL of exonuclease-treated MIP capture reaction to each individual well (*see* **Note 8**).
    (d) Using an RT-PCR thermocycler, amplify the reaction until the reaction begins to plateau under the following conditions: 98 °C for 30 s, followed by 20–25 cycles of 98 °C for 10 s, 60 °C for 30 s, and 72 °C for 30 s (*see* **Note 9**).
12. Standard PCR.
    (a) Prepare a PCR master mix by combining 12.5 μL of 2×iProof High Fidelity Master Mix, 0.125 μL of 100 μM

universal MIP barcode forward primer, and 6.25 μL of nuclease-free water.

(b) Add 18.75 μL of RT-PCR master mix to each well.

(c) Add 1.25 μL of 10 μM individual barcode primers and 5 μL of exonuclease-treated MIP capture reaction into each individual well.

(d) Using a PCR thermocycler, amplify the reaction under the following conditions: 98 °C for 30 s, followed by 20–25 (established in **step 11** of the real-time PCR protocol) cycles of 98 °C for 10 s, 60 °C for 30 s, and 72 °C for 30 s with a final extension time of 72 °C for 2 min and 4 °C forever.

13. Product pooling, clean-up, and gel electrophoresis.

(a) For each plate of DNA samples pool 5 μL of each PCR reaction into a 1.5 mL tube (5 μL×96 = 480 μL) (*see* **Note 3**).

(b) Determine the correct ratio of beads to pooled MIP library by using a bead titration (*see* **Note 10**).

(c) Add 0.9 μL of Agencourt AMPure XP beads per 1 μL of pooled PCR reaction, e.g., (432 μL per 480 μL of pooled PCR reaction). Vortex the tube thoroughly and pulse spin down to remove the beads from within the cap (*see* **Note 11**).

(d) Incubate the sample pool with the beads for 10 min at room temperature.

(e) Place the tube on the DynaMag™-2 magnet, lift the cap and allow the beads to adhere to the side of the tube nearest the magnet for 5 min.

(f) Slowly remove the supernatant using a pipette without disturbing the bead pellet. If the bead pellet is disturbed, pipette them back into the tube and wait a further 1–3 min for the beads to re-bind.

(g) Wash the bead pellet by adding 1 mL of 70% ethanol to fully immerse the beads while the tube is still attached to the magnet. Do not disturb the bead pellet and incubate for 30 s.

(h) Remove the supernatant and repeat step (13 g).

(i) Remove the supernatant completely from the tube, making sure that there is no ethanol left at the bottom of the tube without disturbing the bead pellet (*see* **Note 12**).

(j) Allow the beads to dry for 5 min (*see* **Note 13**).

(k) Remove the tube containing the beads from the magnet and add 100 μL of EB buffer; mix well by manually pipetting up and down at least ten times. Allow the beads to sit at room temperature for 1 min (*see* **Note 14**).

(l) Transfer the tube back to the magnet and incubate for at least 1 min allowing the beads to separate from the EB buffer and adhere to the side of the tube.

(m) Transfer the supernatant, which contains the cleaned MIP library, to a new 1.5 mL tube. Individual MIP libraries can be stored at 4 °C short term or −20 °C for longer periods.

(n) Run the MIP library on a 2% E-Gel® EX Gel by combining 2 μL of pooled MIP library with 18 μL of distilled water and loading 20 μL into the individual wells. Prepare a 100 bp DNA ladder by preparing a 1:1 ratio of E-Gel® Low Range Quantitative DNA Ladder with distilled water and load into the first or final wells in the gel (20 μL).

(o) Run gel electrophoresis for 20 min using the E-Gel® EX Gel platform and confirm the presence of a 276 bp product (*see* **Notes 15** and **16**).

14. Massively parallel sequencing.

Quantitate and Pool MIP Libraries

(a) Prepare individual pooled libraries for sequencing by normalizing each individual library against the concentration of the lowest library within the set pools.

(b) Use the Qubit dsDNA High-Sensitivity assay kit to determine the concentration of each individually barcoded library [9, 10].

(c) Combine each library at equal concentration and determine the final concentration of pooled MIP library as in **step 14.b**.

(d) The size of the MIP megapool, the number of pooled samples, and the desired depth of coverage will determine the individual sequencing requirements. The following protocol uses the Illumina MiSeq platform to test and rebalance individual MIP libraries (*see* **Note 17**).

Denature and Dilute MIP libraries

(e) Denature and dilute MIP libraries according to the Standard Normalization Methods described in the MiSeq Denature and Dilute Libraries Guide [11].

(f) Prepare a fresh 0.2 N dilution of NaOH by combining 200 μL of stock 1 N NaOH and 800 μL of DH_2O.

(g) Dilute the MIP library to 2 nM; then add 5 μL of the library to 5 μL of 0.2 N NaOH.

(h) Vortex the tube thoroughly and pulse spin down to remove the liquid from the lid. Incubate for 5 min at room temperature.

(i) Prepare a 20 pmol denatured library by adding 990 μL of chilled HT1 Buffer to 10 μL of denatured MIP library.

(j) Dilute the denatured 20 pmol MIP library according to desired MiSeq loading concentrations (6–20 pmol). 10 pmol is usually optimal for the majority of MIP libraries.

Loading the MiSeq Reagent Cartridge

(k) Load the diluted MIP library (6–20 pmol) into the MiSeq reagent cartridge according to the MiSeq: Reagent Kit v3-Preparation Guide [12].

(l) Prepare the forward, reverse, and index sequencing primers to a concentration of 10 μM and load into the MiSeq reagent cartridge, according to the MiSeq: Reagent Kit v3-Preparation Guide (*see* **Note 18**) [12].

(m) Set up a sequencing run according to the MiSeq System User Guide [13].

15. Assessment of MIP performance.

(a) Assess capture uniformity by plotting the depth of coverage for individually mapped MIPs.

(b) Normalize read counts for each individual MIP by the total number of reads mapped.

(c) Sort in descending order and plot the ranked uniformity of MIPs in Log_{10} scale.

(d) Rebalance poor-performing MIPs by increasing the relative concentration of MIPs that are one order of magnitude lower in abundance (*see* **Note 19**) (Fig. 2).

(e) Return to methods **step 3.2** and set up the MIP capture using the rebalanced MIP megapool.

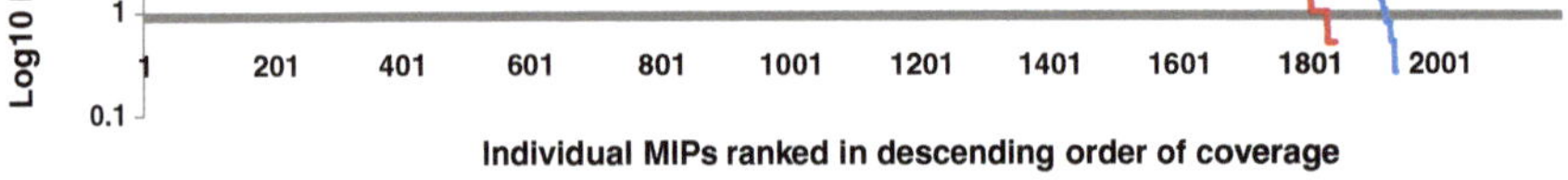

Fig. 2 Capture uniformity for 2196 MIPs "pre" (*blue*) and "post" (*red*) rebalancing. MIPs that perform poorly (one order of magnitude lower in abundance) are rebalanced at a ratio of 50:1 (bad vs. good MIPs) and a substantial number of MIPs are rescued (*green*) upon rebalancing

4 Notes

1. Download the MIPgen design and analysis suite of tools from GitHub (https://github.com/shendurelab/MIPGEN). Use MIPgen to design MIPs across your regions of interest. Note that there are several other dependencies for running this software (e.g., SAMtools, BWA, Tabix) successfully in your local environment.
2. MIPs can be customized to target moderate- and high-complexity DNA targets ranging from 120 to 250 base pairs in size. Low complexity and high GC regions of the genome perform poorly in this assay due primarily to the reliance of the method on PCR amplification and Illumina sequencing. Select your MIPs to be synthesized based on the SVR scores, logistic scores and failure flags (see the MIPGEN README file that accompanies this software package).
3. For ease of handling, use an 8-channel pipette to pool 5 μL of 100 μM MIPs from each well in the 96-well plate. Each tube in the 8-cap strip represents a combined sum of 12 wells or MIPs (5 μL × 12 wells = 60 μL) which can be pooled together to generate a 96 MIP pool containing a volume of 480 μL (60 μL × 8 strip tubes).
4. If the volume of MIP megapool is too small for manual pipetting, dilute the MIP megapool to a lower concentration e.g., 1:1000, so a higher volume can be added. Dilutions should be made fresh for each capture reaction.
5. During this step be sure to create an air-tight seal using the adhesive PCR seal. Use a 10 °C lid offset for each step of the reaction. Failure to perform this step thoroughly will cause the DNA to evaporate during the capture reaction.
6. Capture incubation times may be reduced depending on input DNA concentration. Minimum working DNA stocks should not be less than 100 ng total for the MIP capture reaction.
7. Before adding the exonuclease treatment, cool down the capture plates using cold blocks and prepare the reaction mix on ice. Dispense the exonuclease clean-up mix in equal volumes across a set of 8-Flat-Cap Strip Tubes. Use an 8-channel pipette to dispense 2 μL of exonuclease reaction mix into each capture reaction.
8. RT-PCR is performed using a universal forward primer (MIP_universal_forward: AATGATACGGCGACCACCGAGATC TACACATACGAGATCCGTAATCGGGAAGCTGAAG) and an individual reverse primer (MIP_barcode_reverse: CAAG CAGAAGACGGCATACGAGATNNNNNNNNACAC GCACGATCCGACGGTAGTGT) containing a unique 8mer barcode sequence, which is used for subsequent pooling and sequencing.

9. DNA samples extracted and stored under different conditions will reach plateau at different points during PCR cycling. It is recommended that RT-PCR be performed on a small number of samples representative of each particular sample set so that the correct number of cycles can be established. It is common for a percentage of samples to reach plateau at different cycle points. Select the cycle in which the majority of samples are still within log linear phase before plateau. Once completed, standard PCR may subsequently be performed using the correct number of cycles per sample set.
10. Small contaminants and undesired PCR products are removed during the bead clean-up. However, the ratio of beads to PCR product may vary depending on the size of the MIP library. Here, we use a concentration of 0.9× beads to clean up the pooled MIP library. To determine the quantity of beads to use, perform a bead titration by cleaning up control libraries with varied ratios of beads to MIP library (e.g., 0.8×–1.4× beads) and evaluating by agarose gel electrophoresis. As Agencourt AMPure XP beads preferentially bind to larger DNA fragments, the desired MIP PCR product (276 bp) can be saved while removing other nonspecific PCR products.
11. Allow 100% Agencourt AMPure XP beads to come to room temperature before beginning clean-up. Vortex thoroughly to resuspend the beads into the buffer and dissolve the bead pellet at the bottom of the tube.
12. Tap the magnet gently to consolidate the ethanol at the bottom of the tube and use a p10 pipette tip to remove any residual ethanol.
13. Exceeding 5 min drying time may result in a lower DNA yield.
14. Optimize the amount of elution buffer added to individual MIP libraries to achieve the desired concentration. Smaller MIP pools can typically be eluted in lower volumes.
15. The 276 bp MIP product is specifically based on capturing 162 bp of target sequence using targeting arm lengths of 40–45 bp and single-molecule tags of 5 bp.
16. A small amount of nonspecific product (150 bp) may still remain after bead clean-up; as long as the MIP library (276 bp) represents the predominant band, this should not impact further sequencing steps.
17. Paired-end 101 bp reads are sufficient to sequence individual MIP amplicons of 276 bp, capturing 162 bases with arm lengths of 40–45 bp, 5–8 bp single-molecule tags with enough overlap for read assembly. This can be modified according to the specifics of the individual sequencing library.
18. Sequencing is performed using forward primer: 5′ CATAC GAGATCCGTAATCGGGAAGCTGAAG 3′, MIPseq reverse

primer: 5′ ACACGCACGATCCGACGGTAGTGT 3′, and MIPseqindexprimer: 5′ACACTACCGTCGGATCGTGCGTGT 3′.

19. Poor-performing MIPs can be recovered by "spiking" MIPs in increased relative concentrations, termed rebalancing. Separate MIPs that perform well at 1× concentration from MIPs that require rebalancing. Phosphorylate these MIP pools separately, then pool at a ratio 10:1, 50:1, and 100:1 (poor performers: good performers). MIPs that perform particularly poorly, for example, those that generate zero sequence reads, may not be recoverable and can affect the overall performance of the MIP pool. It is recommended to do a second test run of the rebalanced MIP pool before testing large sample numbers. Check for large proportions of off-target reads indicative of rare MIPs with high off-target capture. This can be avoided by checking output files from the MIPgen design files for MIPs that have over represented arm sequences.

Acknowledgments

We thank Bradley P. Coe for his critical review of the manuscript and Tonia Brown for assistance with the manuscript preparation. We thank Brian J. O'Roak, Beth Martin, Evan A. Boyle, and Joseph B. Hiatt for their overall contributions to developing the MIP protocol. S.C. is supported by a National Health and Medical Research Council (NHMRC) CJ Martin Biomedical Fellowship (#1073726). H.A.S. is supported, in part, by the NHGRI Interdisciplinary Training in Genome Science Grant (T32HG00035). E.E.E. is an investigator of the Howard Hughes Medical Institute. J.S. is an investigator of the Howard Hughes Medical Institute.

Competing financial Interests

E.E.E. is on the scientific advisory board (SAB) of DNAnexus, Inc., and is a consultant for the Kunming University of Science and Technology (KUST) as part of the 1000 China Talent Program.

References

1. Mamanova L, Coffey AJ, Scott CE et al (2010) Target-enrichment strategies for next-generation sequencing. Nat Methods 7:111–118
2. Hardenbol P, Baner J, Jain M et al (2003) Multiplexed genotyping with sequence-tagged molecular inversion probes. Nat Biotechnol 21:673–678
3. Porreca GJ, Zhang K, Li JB et al (2007) Multiplex amplification of large sets of human exons. Nat Methods 4:931–936
4. Turner EH, Lee C, Ng SB et al (2009) Massively parallel exon capture and library-free resequencing across 16 genomes. Nat Methods 6:315–316
5. O'Roak BJ, Vives L, Fu W et al (2012) Multiplex targeted sequencing identifies recurrently mutated genes in autism spectrum disorders. Science 338:1619–1622
6. Boyle EA, O'Roak BJ, Martin BK et al (2014) MIPgen: optimized modeling and design of molecular inversion probes for targeted resequencing. Bioinformatics 30:2670–2672

7. Hiatt JB, Pritchard CC, Salipante SJ et al (2013) Single molecule molecular inversion probes for targeted, high-accuracy detection of low-frequency variation. Genome Res 23:843–854
8. O'Roak BJ, Stessman HA, Boyle EA et al (2014) Recurrent de novo mutations implicate novel genes underlying simplex autism risk. Nat Commun 5(5595):1–6
9. Qubit® Assays: Quick Reference Guide: Pub. no: MAN0010876
10. Qubit® 2.0 Fluorometer: MAN0003231
11. MiSeq: Denature and Dilute Libraries Guide: 15039740v1
12. MiSeq: Reagent Kit v3- Preparation Guide: Part#15044983
13. MiSeq System User Guide: part # 15027617

Chapter 7

Analyzing Copy Number Variation Using Pulsed-Field Gel Electrophoresis: Providing a Genetic Diagnosis for FSHD1

Richard J.L.F. Lemmers

Abstract

The myopathy facioscapulohumeral muscular dystrophy type 1 (FSHD1) is caused by copy number variation of the D4Z4 macrosatellite repeat on chromosome 4. In unaffected individuals the number of 3.3 kb D4Z4 units varies between 8 and 100, whereas 1–10 units are seen in FSHD1 cases. A homologous and heterogenous D4Z4 array can be found on chromosome 10q, but contractions of this array are typically not associated with FSHD. Discriminating between the chromosome 4 and chromosome 10 D4Z4 arrays, as well as determining the array size, requires the use of pulsed-field gel electrophoresis, Southern blotting, and the isolation of high-quality DNA.

Key words FSHD, Macrosatellite repeat, D4Z4, DNA agarose plug, Southern blot, Copy number variation (CNV), Pulsed-field gel electrophoresis (PFGE)

1 Introduction

FSHD is caused by the derepression of the *DUX4* gene, of which a copy is localized in each unit of the D4Z4 macrosatellite repeat array on chromosome 4q [1, 2]. Stable transcription of the *DUX4* gene in somatic tissue, however, requires the presence of an additional polyadenylation sequence containing exon immediately distal to the repeat in the pLAM region [3, 4]. The size of the D4Z4 array determines whether a person is at risk of developing the disease [5]. Each D4Z4 unit in the array is 3.3 kb, and in unaffected individuals the size of the D4Z4 array ranges between 8 and 100 units [6]. In the most common form of FSHD (FSHD1) the array is contracted to a size between 1 and 10 units [7], resulting in a local chromatin relaxation and *DUX4* expression [8, 9]. A linear correlation has been found between the array size and the level of CpG methylation, with shorter arrays being more hypomethylated [10]. In the less common form of FSHD (FSHD2), the disease is mainly caused by mutations in the

Stefan J. White and Stuart Cantsilieris (eds.), *Genotyping: Methods and Protocols*, Methods in Molecular Biology, vol. 1492, DOI 10.1007/978-1-4939-6442-0_7, © Springer Science+Business Media New York 2017

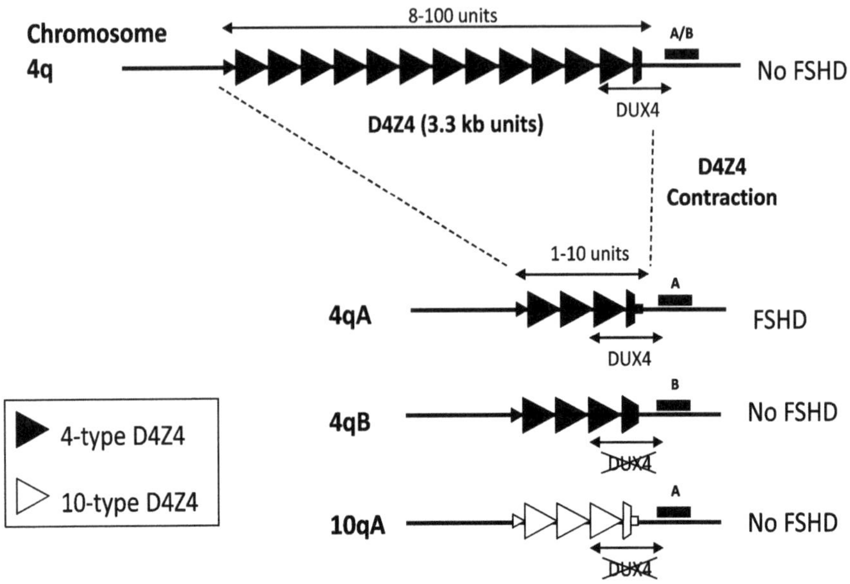

Fig. 1 Schematic overview of the genetic mechanism in FSHD1. The D4Z4 repeat arrays on chromosome 4 (*black triangles*) and the homologue on chromosome 10 (*open triangles*) are depicted as *sequential triangles* (units). The presence or absence of the complete *DUX4* gene is indicated, as is the position of the 4qA or 4qB probes. In unaffected individuals the size of the D4Z4 repeat array ranges between 8 and 100 units, whereas patients with FSHD1 have a D4Z4 repeat array size between 1 and 10 units (on 4qA, but not on 4qB or 10q)

SMCHD1 gene on chromosome 18 [11]. The D4Z4 array size is still important, as the majority of FSHD2 patients have an array that ranges between 8 and 16 units [10].

Genetic analyses in FSHD are complicated by the genetic variation on chromosome 4q. Two almost equally common forms have been described, 4qA and 4qB [12]. The 4qB form is not associated with FSHD, as it lacks the important additional exon distal to the repeat containing the *DUX4* polyadenylation sequence [3, 4]. Furthermore, a homologous D4Z4 array can be found on chromosome 10q, which also ranges between 1 and 100 units and is homologous to 4qA. However, derepression of D4Z4 on chromosome 10 by either an FSHD1 or an FSHD2 mechanism does not result in stable *DUX4* expression, due to a mutation in the *DUX4* polyadenylation sequence (Fig. 1) [4, 13].

The genetic diagnosis of FSHD is further complicated due to rearrangements between chromosome 4-type and 10-type D4Z4 arrays. These result in arrays consisting of a mix of both 4q and 10q D4Z4 units, or in complete 4-type D4Z4 arrays on chromosome 10 [14]. Remnants of D4Z4 evolution give rise to complex D4Z4 profiles in at least 30% of African, European, and Asian individuals [15]. In about 10% of cases, FSHD is caused by a new rearrangement of which half are gonosomal mosaic for the repeat contraction [16, 17]. To estimate the recurrence risk, it is important to identify these gonosomal mosaic cases. The genetic

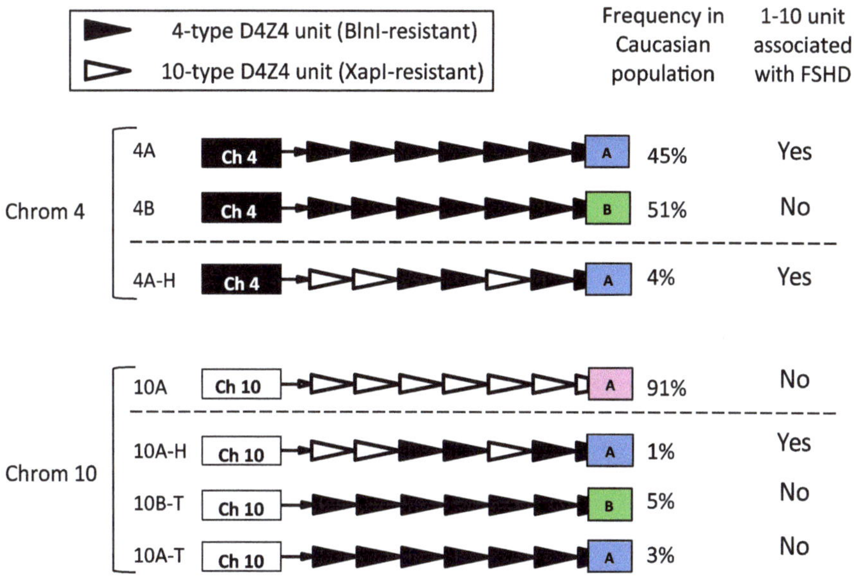

Fig. 2 The genetic analysis of FSHD is complicated by the presence of complex D4Z4 alleles. These alleles are remnants of D4Z4 evolution, and can be found on chromosome 4 (4A-H) and on chromosome 10 (10A-H, 10B-T, and 10A-T) [15]. The frequency of these alleles in the general Western European population is indicated. Repeat contractions on 4A, 4A-H, and 10A-H cause FSHD, and other allele types are not associated with FSHD

mechanism necessitates the discrimination between 4qA, 4qB, and 10q D4Z4 arrays. This should be performed on high-quality DNA, to allow the visualization of all D4Z4 fragments that can be more than 300 kb in size. The most commonly used diagnostic method is Southern blotting of genomic DNA after digestion with a specific set of restriction enzymes, and subsequent hybridization with haplotype-specific probes [18, 19]. Ideally this method is performed by using pulsed-field gel electrophoresis (PFGE) in combination with high-quality DNA [20] (Fig. 2). Recently an alternative method has been developed based on in situ hybridization of stretched DNA [21]. As Southern blot-based FSHD diagnostics is currently the gold standard, this chapter focuses on this method.

2 Materials

2.1 Isolation of Cells

1. EDTA blood tube (9 mL) or lithium heparin blood tube (9 mL) (*see* **Note 1**).
2. Erythrocyte lysis buffer: 155 mM NH_4Cl, 10 mM $KHCO_3$, 1 mM EDTA pH 8.0.
3. Beckman centrifuge (type GS-6R).
4. Laminar flow safety hood.

2.2 Preparation of Agarose Blocks

1. Perspex mold for blocks (volume 100 μL): Dimensions of half block slightly smaller than dimensions of the agarose gel wells.
2. 1000 μL Pipet (Gilson/Eppendorf).
3. 200 μL Pipet (Gilson/Eppendorf).
4. SE buffer: 75 mM NaCl, 25 mM EDTA pH 8.0.
5. SE with 1.4 % InCert agarose dissolved and store at 20 °C (*see* **Note 2**).
6. InCert agarose (FMC BioProducts)

   SE with 1 % *N*-lauroyl-sarcosine (sarcosyl).
7. Pronase (20 mg/mL): Dissolve pronase in 10 mM NaCl/10 mM Tris–HCl (pH 7.5) to a final concentration of 20 mg/mL. Incubate for 1 h at 37 °C. Store at –20 °C.
8. Water bath 37 °C.
9. 0.5 M EDTA pH 8.0.

2.3 Equilibration Agarose Blocks and Treatment with Endonuclease

1. TE^{-4} Solution: 10 mM Tris–HCl pH 7.4, 0.1 mM EDTA.
2. Restriction enzymes:

   EcoRI

   HindIII

   BlnI (AvrII)

   XapI (ApoI)
3. Appropriate restriction enzyme buffers:

   Buffer B (EcoRI and HindIII double digest), stock 10×

   Buffer H (EcoRI and BlnI double digest), stock 10×

   Buffer Y Tango (XapI digest), stock 10×

   Add 0.1 M Spermidine to a final concentration in digestion of 3.3 mM

   Add 1 M DTT to a final concentration in digestion of 1 mM

2.4 Pulsed-Field Gel Electrophoresis

1. 10× TBE buffer: 890 mM Tris pH 7.6, 890 mM boric acid, 20 mM EDTA.
2. PFGE-suitable agarose (e.g., multiple-purpose agarose; Roche).
3. 10 mg/mL Ethidium bromide solution.
4. PFGE apparatus (CHEF DRII System, Biorad).
5. Cooling module.
6. Molecular weight marker for PFGE.

2.5 Southern Blotting

1. UV transilluminator.
2. Platform shaker.
3. UV Stratalinker (Stratagene 1800).

4. Whatman GB003 gel blotting paper (20 × 20cm and 58 × 60cm).
5. Hybond XL: charged nylon membrane (GE Healthcare Amersham).
6. Cellulose sheets.
7. Blotting buffer: 0.4 M NaOH, 0.6 M NaCl (alternatively 0.4 M NaOH).
8. Neutralizing buffer: 2× SSC (SSC buffer 20× contains 3 M NaCl in 0.3 M sodium citrate, pH 7.0), 0.2 M Tris–HCl pH 7.5.

2.6 Prehybridization and Hybridization

1. NaPi/PEG hybridization buffer: 125 mM $NaHPO_4$ pH 7.2, 10% polyethylene glycol 6000, 0.25 M NaCl, 1 mM EDTA, 7% SDS.
2. Fish sperm DNA (10 mg/mL).
3. Water bath 65 °C, or hybridization oven.
4. Heat block 37 °C.
5. Heat block 95 °C.
6. Megaprime labeling kit (Amersham).
7. DNA probes (p13E-11, 4qA, and 4qB).
8. Washing buffers:
   (a) Wash buffer 1: 2× SSC and 0.1% SDS
   (b) Wash buffer 2: 1× SSC and 0.1% SDS
   (c) Wash buffer 3: 0.3× SSC and 0.1% SDS
   (d) Wash buffer 4: 0.1× SSC and 0.1% SDS

3 Methods

The genetic analysis of all FSHD cases requires high-quality DNA. The common methods of DNA isolation, either manually or in an automatic system, result in liquid DNA. However, mechanical stress and the ethanol precipitation step during the preparation of liquid DNA generally reduce genomic DNA fragments to <100 kb, and further shearing during the handling of the DNA may result in even smaller fragments. Liquid DNA is therefore not generally suitable to size D4Z4 repeat arrays larger than 50 kb, and not all D4Z4 arrays from 4q and 10q in an individual might be visualized. Agarose embedded DNA blocks are required to visualize all D4Z4 array repeats by Southern blotting. DNA shearing is prevented as the cells are embedded in agarose prior to the pronase and detergent (Sarkosyl) treatment, and further DNA isolation steps are not needed.

The DNA sequence of the different D4Z4 arrays from chromosomes 4 and 10 shares 99% similarity. Chromosome 4- and

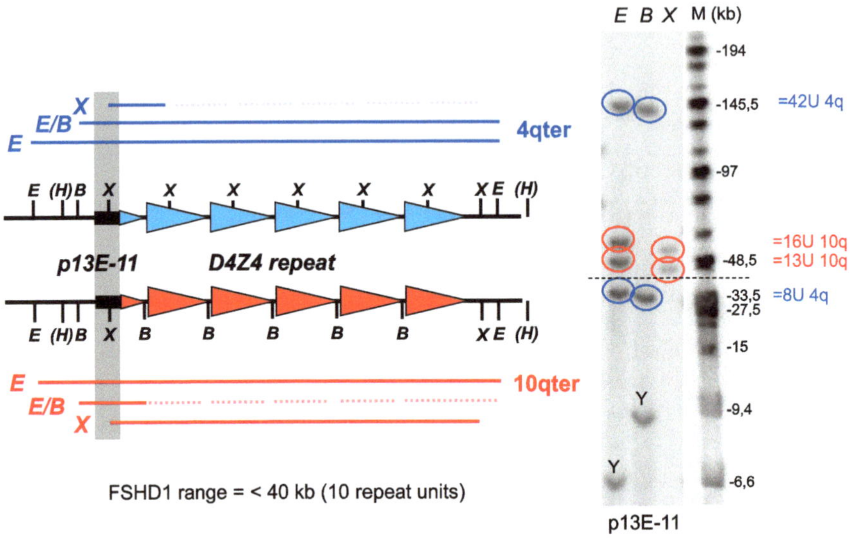

Fig. 3 Schematic overview of Southern blot analysis of the FSHD locus by different restriction enzymes. The D4Z4 repeat array on chromosome 4q35 and 10q26 is highly polymorphic and ranges in size from 1 to 100 repeat units (chromosome 4 and 10 D4Z4 units depicted as *blue* and *red*, respectively). After restriction enzyme digestion and electrophoresis, genomic DNA is transferred to a nylon membrane by Southern blotting and then hybridized with probe p13E-11. The EcoRI (or alternatively EcoRI and HindIII double) digestion (E) reveals all four repeat arrays derived from chromosomes 4 (*blue*) and 10 (*red*). Chromosome 10-type D4Z4 repeat arrays carry a BlnI site in each unit and are fragmented upon EcoRI and BlnI double digestion (B). Similarly, chromosome 4q repeat arrays are fragmented by the XapI restriction enzyme (X). Consequently, digestion B and X only shows the chromosome 4-type or chromosome 10-type fragments, respectively. The unit size of the different chromosome 4 and 10 alleles is calculated based on **Note 3**. As the XapI restriction enzyme also digests within the p13E-11 probe region, the fragments in lane (X) are weaker than in lanes (E) and (B). Probe p13E-11 also recognizes a Y chromosome fragment of approximately 6.6 kb upon EcoRI and HindIII double digestion, and 9.4 kb upon EcoRI or EcoRI and BlnI double digestion. The marker lane (M) is indicated

10-specific restriction enzyme recognition sites in the different D4Z4 sequences are used to discriminate between both repeat arrays. The BlnI (or AvrII) restriction site can only be found within each chromosome 10-type D4Z4 unit, whereas the XapI (ApoI) restriction site is specific for chromosome 4-type D4Z4 units [18, 19]. To release the complete D4Z4 repeat arrays on chromosomes 4q and 10q, genomic DNA is digested with restriction enzymes (in general with EcoRI or with EcoRI and HindIII) that cut at both ends of the D4Z4 repeat array [7, 18, 22]. To observe chromosome 4q-type D4Z4 arrays a double digestion with EcoRI and BlnI is performed, whereas a restriction enzyme digestion with XapI only shows the 10q-type arrays (Fig. 3). Both the EcoRI and EcoRI/HindIII D4Z4 fragments have sequence flanking the D4Z4 repeat array. These sequences have to be considered when calculating the number of D4Z4 units per array (*see* **Note 3**). For each of these reactions approximately 500,000 cells (approximately 3.5 micrograms DNA) in a half agarose block are required, or

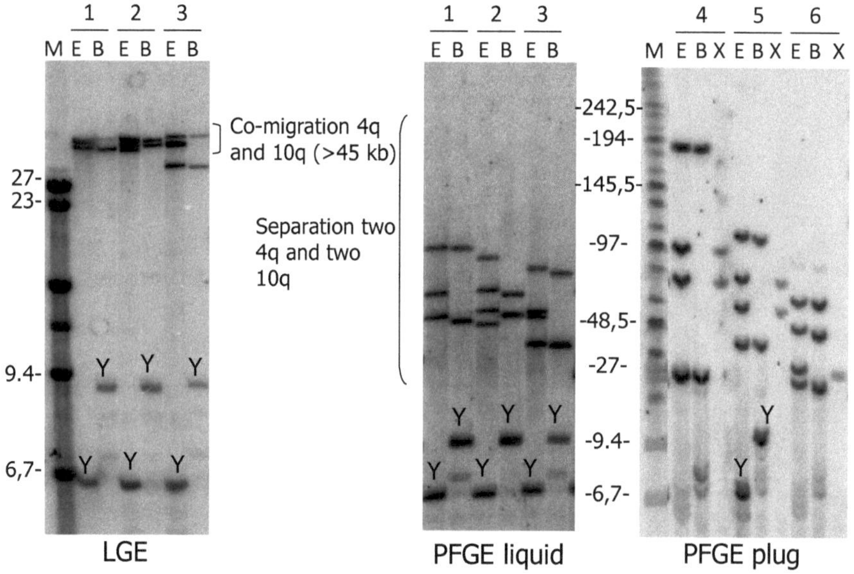

Fig. 4 Southern blot analysis of DNA separated by LGE and PFGE showing the superiority of PFGE in the separation of large-size fragments. Liquid DNA samples 1, 2, and 3 have been analyzed by both LGE and PFGE after digestion with EcoRI/HindIII (E) and EcoRI/BlnI (B). As indicated the LGE result shows co-migration of the largest 4q and 10q fragments, while these fragments are separated on the PFGE gel. The liquid DNA in this example has a high quality, but one of the chromosome 10 fragments of individual 1 is not visible due to DNA shearing. On the right, a PFGE gel is depicted using DNA agarose blocks. DNA on this gel is digested with EcoRI/HindIII (E), EcoRI/BlnI (B), and XapI (X). The superiority of block DNA above liquid DNA is clearly noticeable as fragments up to almost 300 kb are visible after hybridization. All three Southern blots have been hybridized with probe p13E-11. The marker lanes (M) are indicated. Samples 1, 2, 3, and 5 are from males, and the co-hybridizing chromosome Y fragment is indicated (Y)

alternatively 5 micrograms of liquid genomic DNA. Digested DNA is separated by pulsed-field gel electrophoresis (PFGE). PFGE together with agarose-embedded DNA blocks allows separation of fragments up to hundreds of kilobases [22]. In contrast, linear gel electrophoresis (LGE) allows the sizing of DNA fragments between 3 and 50 kb, and therefore only enables identification of FSHD1 with standard 4q-type arrays between 1 and 10 D4Z4 units. Figure 4 illustrates the differences between high quality liquid DNA and DNA embedded in agarose plugs, and between PFGE and LGE.

After restriction enzyme digestion and electrophoresis, genomic DNA is transferred to a nylon membrane by Southern blotting and hybridized with probe p13E-11 [7]. Probe p13E-11 recognizes the region immediately proximal to D4Z4 contained within the EcoRI fragment, and allows the chromosomal origin to be determined in most cases. To minimize nonspecific hybridization of probe to the blot, a prehybridization or blocking step is required. Salmon sperm DNA is commonly used as a blocking agent. For estimating the size of the array, a high-molecular-weight marker (MWM) is used (often based on phage lambda DNA). Southern blot hybridizations are often performed with radioactive-labeled probes using the isotope phosphorus-32 (^{32}P). Alternatively, non-radioactive probe labeling can be applied.

To determine the D4Z4 repeat array genotype a similar approach is applied, using the restriction enzyme HindIII and probes 4qA and 4qB. This can be hybridized on the same Southern blot [12]. D4Z4 repeat arrays on chromosome 10 are mainly of the A-type, while D4Z4 arrays on chromosome 4 can be either A or B (Fig. 2). Combining information from p13E-11 Southern blot (repeat size and chromosomal origin) and the 4qA/4qB blots will provide the genotype (Fig. 5).

More detailed FSHD genotyping can be performed. A simple sequence length polymorphism (SSLP) is located 3 kb proximal to the D4Z4 array [23]. For FSHD2, the uncommon form of FSHD, the mutations in *trans* in the gene encoding the epigenetic modifier SMCHD1 are associated with D4Z4 hypomethylation on chromosomes 4q and 10q, which can be detected by methylation-specific methods (Southern blots or bisulfite sequencing methods). These methods are not discussed here, but have been described elsewhere [11, 24–26].

3.1 Cell Isolation and the Preparation of Single-Cell Suspension for Agarose Blocks

DNA source: White blood cells (leucocytes) isolated from whole blood or any cultured cells (lymphoblastoid, fibroblast, or myoblast cell lines).

For white blood cells continue from **step 1**; for cultured cells trypsinize cells for adherent cultures, count cells, and then continue with **step 5**.

1. Collect 5–10 mL of whole blood in EDTA tube and keep it at 20 °C for at least 2 days (and maximum 7 days) (*see* **Notes 4** and **5**).
2. Transfer blood to 50 mL tube, add 25 mL of Erythrocyte lysis buffer, and put the tube on ice to lyse red blood cells. This takes 5–8 min, with the solution turning very dark red.
3. Centrifuge cells at 20 °C for 8 min at 266 × *g* (brake high).
4. Aspirate the supernatant, resuspend the pellet in 15 mL Erythrocyte lysis buffer, and transfer cell suspension to 15 mL conical tube (*see* **Note 6**).
5. Centrifuge cells at 20 °C for 5 min at 266 × *g* (brake high).
6. During the centrifugation, melt the SE/1.4% agarose buffer and place the tube at 60 °C in a water bath to prevent the agarose setting.
7. Put tape on one side of the plastic block mold and place mold on ice.
8. Estimate the number of cells based on the size of the white blood cell pellet (using reference tube, *see* **Note 7**), or for cul-

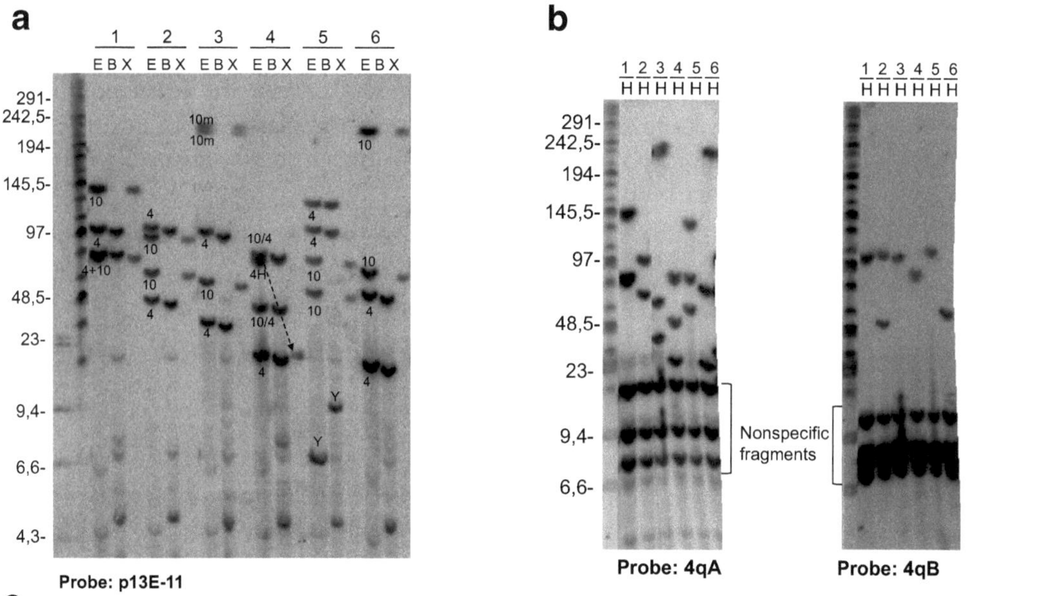

c

		Allele 4_1			Allele 4_2			Allele 10_1			Allele 10_2			
sampleID	M/F	units	kb	A/B	units	kb	A/B	units	kb	A/B	units	kb	A/B	Result
1	F	22	79	A	29	99	B	22	79	A	41	141	A	No FSHD1
2	F	12	45	B	29	99	B	19	68	A	27	94	A	No FSHD1
3	F	9	35	A	28	98	B	17	62	A	[65(50%);67(50%)]	[219(50%);225(50%)]	A	FSHD1
4	F	3	15	A	21	73H1	A	10	39(10/4)	A	22(10/4)	78(10/4)	B	FSHD1
5	M	29	101	B	37	128	A	15	54	A	22	79	A	No FSHD1
6	F	3	15	A	14	52	B	20	71	A	65	218	A	FSHD1

Fig. 5 (**a**) PFGE blot hybridized with p13E-11, as described in Fig. 3. Cross-hybridizing chromosome Y fragments are indicated (Y). Sample 3 carries a mosaic chromosome 10q fragment (10 m). Individual 4 carries a hybrid chromosome 4q fragment (4H) and two translocated 4q-like repeats on chromosome 10 (10/4). (**b**) 4qA and 4qB hybridizations of HindIII digested DNA from the same individuals as shown left. Indicated are the a-specific fragments in the region between 15 kb and 7 kb (4qA) and 12 kb and 7 kb (4qB). Most chromosome 10q fragments carry the distal 4qA variation, except for one translocated chromosome 10q variant (haplotype 10B161T, *see* Fig. 2). (**c**) Interpretation of genotypes after Southern blot hybridizations in (**a**) and (**b**). Individuals 1, 2, and 5 are determined as "No FSHD1." Individuals 3, 4, and 6 carry a short repeat array (9, 3, and 3 units, respectively) on a 4qA chromosome and are determined as "FSHD1." The unit size of the different alleles is calculated as described in **Note 3**

tured cells use the exact number of cells (counted prior to the centrifugation).

9. Add the calculated volume of SE (about 1500 μL for 10 mL of blood) to the pellet to a concentration of approximately 20 million cells per mL in SE, and resuspend the cells using a 10 mL tip (*see* **Note 8**).
10. Add an equal volume of the 60 °C SE/1.4% agarose solution and resuspend gently.
11. Dispense mixture over the wells in the plastic mold with the 1000 uL tip and leave the blocks to set.

12. Prepare a 10 mL solution of SE/sarcosyl (1%) with 300 μL pronase in a 15 mL tube.
13. When set, remove the surplus of agarose from the mold using a knife. Push the blocks out of the plastic mold into the SE/sarcosyl/pronase solution, using air pressure from a 1 mL rubber bulb.
14. Incubate the blocks in a 37 °C water bath for at least 2 days.
15. At day three, discard the SE/sarcosyl/pronase solution and wash the blocks with 10 mL water.
16. Discard the water, and add 10 mL 0.5 M EDTA (storage buffer) to the blocks.
17. Store the block DNA at 4 ° C.

3.2 Equilibration of Agarose Blocks and Treatment with Endonuclease

1. Remove the blocks needed for the restriction enzyme digestion from the 0.5 M EDTA buffer. Cut into two with knife, place each half block in a 1.5 mL tube, and add 1 mL water.
2. Remove water from tube without damaging block and add 1 mL TE^{-4}. Rotate 360° for 1–2 h in cold room or at 20 °C.
3. Remove TE^{-4} without damaging block and add 1 mL TE^{-4}. Rotate 360° for 1–2 h in cold room or at 20 °C.
4. Remove TE^{-4} without damaging block and add 1 mL digestion buffer (with spermidine and DTT, without the restriction enzyme). Rotate 360° for 1–2 h in cold room or at 20 °C (*see* **Note 9**).
5. Remove digestion buffer and add 150 μL digestion buffer with restriction enzyme and spermidine and DTT. Digest for 6 h or overnight at 37 °C.

3.3 Pulsed-Field Gel Electrophoresis

1. Prepare 2.5 L electrophoresis buffer and 0.5× TBE with ethidium bromide to a final concentration of 15 μg/L. Remove old running buffer from the PFGE chamber and replace with about 2.5 L of the freshly prepared buffer.
2. Prepare agarose gel with 0.5× TBE and 0.88% PFGE-suitable agarose. Add ethidium bromide to a final concentration of 15 μg/L. In this example a 20×20 cm gel is prepared and these dimensions are also used for the Southern blotting protocol.
3. After solidification of the gel, place a little water on the gel in front and in back of the comb. Keeping the gel in place carefully and gradually (without stopping) remove comb and put water in the wells (do not yet place the gel in the PFGE chamber, loading of DNA blocks is performed on the lab bench).
4. Slide a dark laminated piece of film under the wells of the agarose gel to increase the contrast so that you can see the wells, and distinguish full from empty. Remove digested

(half) block with spatula from tube. Position the end of the spatula in the well and gently push the plug with the end of a 200 μL pipet tip or with your gloved finger, so that it is below the surface of the gel.

5. Add the MWMs (*see* **Note 10**). For accurate fragment sizing, the MWMs should be placed as the first and last samples on the agarose gel.
6. Put gel in the PFGE chamber. Add more electrophoresis buffer so that the gel is ~1 cm below the buffer level.
7. The settings for electrophoresis recommended for the BioRad Chef II are as follows. DNA fragments between 3 and 400 kb: 1 s as start and 20 s as stop time at 21.5 C (two identical cycles of 10 h).DNA fragments between 3 and 80 kb: 1 s as start and 3 s as stop time at 21.5 C (two identical cycles of 10 h) (*see* **Note 11**).

3.4 Southern Blotting

1. Cut three 20 × 20 cm pieces (size agarose gel) and two 20 × 30 cm pieces (paper bridge) of blotting paper (*see* **Note 12**).
2. Prepare the blotting towels (40 × 60 cm cellulose sheets) by folding them so that they are the size of the gel and each individual folded bundle is approximately 1 cm thick. The total layer of folded sheets should be about 12 cm.
3. Cut a 20 × 20 cm (agarose gel size) piece of the Nylon membrane. Label the bottom of the membrane with a waterproof pen. Prior to use it should be briefly pre-soaked in water, and then for 5 min in blotting buffer.
4. Visualize and photograph the DNA on the PFGE agarose gel using an UV transilluminator (wavelength 312 nm). Recommended UV irradiation approximately 1 min to induce single-strand DNA breaks that enable transfer of fragmented DNA onto membranes (excessive irradiation and shorter UV wavelength might fragment the DNA too much, and should be avoided). Alternatively, irradiate the gel with a Stratalinker.
5. Cover the agarose gel with the other transfer plate and rotate the gel 180° so that the bottom of the gel is now up.
6. Place agarose gel in a basin and add blotting buffer until the gel is covered. Gently shake the gel for 15 min. Refresh the blotting buffer and shake for another 15 min. During this 30-min saturation period prepare the blotting setup so that it is ready immediately after the last 15-min saturation step.
7. Gel placement Fig. 6).

   (a) Build up the bridge: Pre-wet the double-thickness 20 × 30 cm bridge in blotting buffer and gently, but quickly, put it on a 20 × 20 cm plate on top of the small

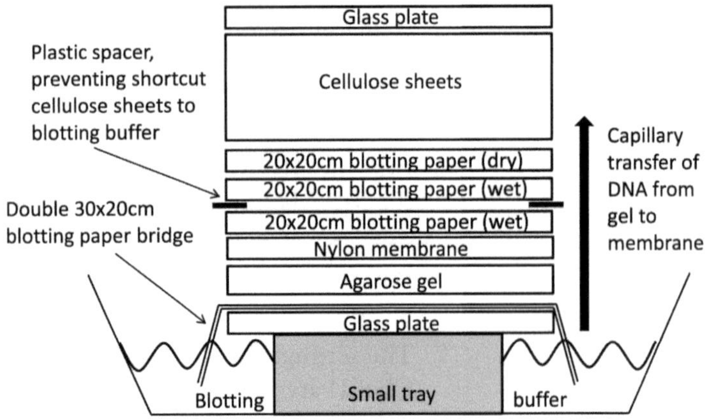

Fig. 6 Assembly for Southern blotting

tray. Immediately lower the flaps of the bridge gently into the blotting buffer and continue with step (b).

(b) Immediately after completion of the last 15-min blotting buffer saturation step, slide the gel onto the bridge; the bottom of the gel should be facing up.

(c) Gently lower the Nylon membrane on the agarose gel.

(d) Pre-wet the first blotting paper sheet in blotting buffer, and slowly lower so that it exactly covers the membrane from one end to another. Put four strips of plastic to make a frame on the wet blotting paper (see Note 13).

(e) Pre-wet the second blotting sheet and lower it exactly into place. Roll out air bubbles at both ends by using a (shortened) plastic 10 mL pipet, starting from the middle toward the end of the membrane/gel.

(f) The third and last piece of blotting paper is placed on dry. Roll out air bubbles as above and keep rolling until it is completely wet.

(g) Put the first of the folded towels on top, followed by the remainder to create an ~12 cm height of towel stack

(h) Add more NaOH/NaCl until the reservoir is filled completely.

(i) Cover paper towel with a plastic plate and a 100 gram weight and leave for at least 6 h or overnight.

8. Membrane release:

(a) After blotting, transfer membrane to 2× SSC/0.2 M Tris–HCl (pH7) neutralizing solution for 5 min.

(b) Dry the membrane shortly between clean filter paper, prior to UV cross-linking. Cross-link DNA to membrane at 120 mJ/cm2 (setting "auto cross-link" in UV Stratalinker 1800, Stratagene). Start hybridization procedure, or store the dry membrane at 20 °C in a dark place (*see* **Note 14**).

3.5 Prehybridization and Hybridization

(Hybridizations can be performed in either a hybridization oven or in a water bath.)

3.5.1 Hybridization in Hybridization Oven

1. Prewarm the NaPi/PEG hybridization buffer in a 65 °C water bath, and heat the salmon sperm DNA to 95 °C for 5 min.
2. Roll the dry Nylon membrane in a 250 mL glass cylinder. Rinse the membrane with washing buffer 3 (0.3× SSC and 0.1% SDS) at room temperature, and ensure that the membrane sticks completely to the glass wall without air pockets between membrane and glass.
3. Remove the washing buffer and add 20 mL prewarmed NaPi/PEG hybridization buffer supplemented with 200 μL salmon sperm DNA to a final concentration of 50 μg/mL.
4. Prehybridize for at least 1 h at 65 °C.
5. During the prehybridization step prepare the probe (*see* Subheading 3.6).
6. After prehybridization add boiled probe to NaPi/PEG (pre-) hybridization buffer (refreshing of pre-hybridization buffer is not necessary).
7. Hybridize for 1–2 nights at 65 °C.

3.5.2 Hybridization in Water Bath

1. Cut the 20×20 cm membrane into two half pieces parallel to the DNA lanes to fit in a 11×21 cm plastic tray (*see* **Note 15**).
2. Prewarm the NaPi/PEG hybridization buffer in a 65 °C water bath. Heat the salmon sperm DNA to 95 °C for 5 min.
3. Add 70 mL of prewarmed NaPi/PEG hybridization buffer supplemented with 700 μL salmon sperm DNA to a final concentration of 50 μg/mL to the plastic tray.
4. Add the sliced membranes one by one. After each slice make sure that the membrane is completely covered by hybridization buffer, without air bubbles in between.
5. Alternatively, two membranes can by hybridized simultaneously. In this case use 90 mL hybridization buffer (supplemented with 900 μL salmon sperm DNA) for the four half-membranes.
6. Prehybridize for at least 1 h at 65 °C by gently shaking. Ensure that the membrane slices are not sticking together.

7. After prehybridization briefly remove the membranes from the prehybridization buffer (keep them in your hand), add the boiled probe to the buffer (refreshing of prehybridization buffer is not necessary), and homogenize the probe/buffer mix. Place back the membrane slices one by one, making sure that they are completely covered by the probe/buffer combination without air bubbles in between.
8. Hybridize overnight at 65 °C in a shaking water bath.

3.6 Preparation Isotope-Labeled DNA Probe

1. DNA probe template can be isolated from either plasmid DNA or by insert PCR.
2. For isolation from plasmid DNA, digest the plasmid with the appropriate restriction enzymes to isolate the plasmid insert. After gel electrophoresis and gel purification, the insert DNA can be used in the labeling reaction.
3. For a plasmid insert-PCR use primers flanking the insert (mostly M13 primers) and 1–10 pg of the plasmid DNA. After gel electrophoresis and gel purification, the purified PCR product can be used in the labeling reaction (*see* **Note 16**).
4. Approximately 20 ng of the purified probe DNA (for both plasmid restriction or insert PCR) can be used in a random primed labeling reaction.
5. Prepare the radioactive probe according to the manufacturer's instructions.
6. Use 1.5 μL ^{32}P-dCTP (activity 3000 Ci/mmol or 111TBq/mmol) per labeling reaction.
7. Prepare one labeled probe for each blot to be hybridized.
8. Optionally, the MWM can be hybridized in the same hybridization reaction. For this, separately label ~ 20 ng of the purified MWM according to the manufacturer's instructions (lambda DNA is provided as standard DNA in the Megaprime kit from Amersham). Use only 1/500 dilution of this MWM probe in your hybridization in combination with your specific probe.
9. The labeling reaction takes 20 min at 37 °C.
10. After labeling, add 80 μL TE^{-4} to the specific probe and 500 μL to the MWM probe.
11. Add 1 μL of the freshly made, diluted MWM probe to the specific probe (*see* **Note 17**).
12. Denaturate probe mixture for 7 min at 95 °C.
13. After denaturation, immediately chill the probe on ice and add the probe to the blots and the hybridization buffer as described in Subheading 3.5 (*see* **Note 18**).

3.7 Washing Blots

1. Prewarm wash buffers in a 65 °C water bath.
2. Discard the hybridization buffer with the radioactive probe(s) from the blots.
3. Rinse the blots with wash buffer 1 and briefly shake and remove the buffer.
4. Add sufficient wash buffer 1 (usually about 100 mL) to ensure that the blots are not sticking together and wash for 15 min at 65 °C in a shaking water bath.
5. Discard the wash buffer and repeat the wash step two times.
6. Check the blot for radioactive signal using a Geiger counter. Readings of 8–30 counts per second (cps) at the position on the blot where you expect to find the labeled DNA are desirable. For background, check the region on the edges of the blot. If the background is too high, wash the blot again with a more stringent wash solution (wash buffer 2, 3, and 4) (*see* **Note 19**).
7. Remove the last washing buffer and dry the blot between Whatman paper.
8. Wrap the blots in Saran Wrap and expose for 1–2 days to a Phosphor Imaging cassette, or for 1 week (in the darkroom) to an X-ray film in a autoradiogram cassette.
9. The radioactive signal can be visualized by scanning the phosphor image screen, or developing the film.
10. For sequential hybridization of the same blot, it is recommended to remove the previous probe signal. In order to do this the probe can be stripped (removed) by adding boiling wash buffer 4 solution to the blot. Shake for several minutes at room temperature and then discard the buffer. Alternatively the blot can be soaked for 30 min in 0.2 M NaOH solution at 45 °C.

4 Notes

1. Blood should preferably be collected in an EDTA tube. When blood is collected in a heparin tube, the isolated white blood cells are easier to resuspend and appear fine. The DNA yield is significantly lower when DNA isolation occurs more than 7 days after blood collection (see **Note 4**).
2. SE with 1.4% InCert agarose can be stored at 20 °C (solid condition). To dissolve, briefly heat in a microwave.
3. In FSHD diagnostics, the size of the FSHD allele is often indicated as the D4Z4 fragment after EcoRI or EcoRI/HindIII, or even after EcoRI/BlnI restriction. To prevent confusion, it is better to assign the number of repeat units in the D4Z4

Table 1 D4Z4 fragment size to repeat unit conversion table. Columns 2, 3 and 4 depicts the size of the Southern blot D4Z4 fragment upon DNA digestion using different restriction enzymes and the first column indicates the corresponding number of units for each combination

Size 4qA D4Z4 fragment (kb)			
Units	**EcoRI**	**EcoRI/HindIII**	**EcoRI/BlnI**
1U	10.2	8.1	7.0
2U	13.5	11.4	10.3
3U	16.8	14.7	13.6
4U	20.1	18.0	16.9
5U	23.4	21.3	20.2
6U	26.7	24.6	23.5
7U	30.0	27.9	26.8
8U	33.3	31.2	30.1
9U	36.6	34.5	33.4
10U	39.9	37.8	36.7
11U	43.2	41.4	40.0
12U	46.5	44.4	43.3
13U	49.8	47.7	46.6

array based on the array size. EcoRI, EcoRI/HindIII, and EcoRI/BlnI D4Z4 fragments all have different sizes flanking the D4Z4 repeat array. For the calculation of the number of D4Z4 units in EcoRI fragments, subtract 6.9 kb and then divide by 3.3 kb. In EcoRI/HindIII fragments; subtract 4.8 kb and then divide by 3.3 kb. For EcoRI/BlnI fragments, subtract 3.7 kb and then divide by 3.3 kb (*see* Table 1).

4. Preferably isolate white blood cells (WBC) between 2 and 7 days after collection. If erythrocyte lysis is started within 2 days after collection the erythrocytes are too fresh and the lysis fails. If the method is applied more than 7 days after drawing the WBC become too fragile and they might lyse while using the erythrocyte lysis buffer, and the remaining white blood cells tend to stick together after pelleting hampering the resuspension of the cells.

5. When blood samples are shipped by plane, temperature changes during the shipment might damage the cells, causing lower yield and difficulties with resuspension. Prevent these problems by isolating the blood tubes in a small styrofoam box.

6. PBLs or cultured cells used for the generation of agarose blocks need to be in a single-cell suspension prior to the agarose

embedment. Avoid cell clumps in the suspension as this will often result in incomplete endonuclease reaction in the DNA agarose plug. Do not leave the cells pelleted, but resuspend as quickly as possible to prevent agglomeration. For efficient resuspension first use a small volume (3–5 mL buffer), and after resuspension increase to desired volume.

7. Use a black marker pen on the tube to indicate the average WBC pellet size after erythrocyte lysis, and use this as reference for estimating the number of cells. On average, 10 mL blood contains 25 million white blood cells.
8. Calculation of DNA concentration per agarose plug. The initial cell suspension has a concentration of ~20 million cells per mL, which will be halved after adding of an equal volume of agarose solution. A whole agarose plug has a volume of 100 μL, or ~1 million cells. There is ~6.6 pg DNA in a human cell. For a single restriction enzyme digestion a half block is required (500,000 cells), which is ~3.3 μg genomic DNA.
9. Do not rotate the plugs over the weekend because the block might be damaged. After equilibration, plugs can be stored at 4 °C for at least 4 weeks.
10. For standard PFGE condition we use the following MWMs: 50 ng/lane liquid lambda DNA solution digested with restriction enzyme HindIII in bromophenol blue dye and concatemerized lambda in agarose block. The fragments of these markers do not overlap, and therefore these MWMs can be added to the same slot. Use the Biorad CHEF DNA Size Standard (8.3–48.5 kb) for run conditions that are focused on fragments below 50 kb.
11. The temperature of the buffer correlates with the migration of the DNA in the gel. A lower temperature makes the run slower. When the temperature is too high, the current of the CHEF device will be interrupted and the displays give an error.
12. Best blotting is achieved using thick blotting paper. We have obtained more variable results when using thinner paper.
13. The plastic strips prevent a possible shortcut between the buffer and the blotting paper and cellulose sheets on top of the membrane. Alternatively, the four plastic strips can be replaced by a plastic overhead transparency sheet, where a square slightly smaller than the gel size is cut out.
14. When storing the blots, make sure that the gels are dried between blotting paper. Preferably start hybridization immediately after blotting. Storage of longer than 2 days can have a negative effect on the blot.
15. With a pen, make a zig-zig (like 2 connected letter "S"s) with a pen parallel to the lanes so that you cover several lanes in the middle of the membrane with the zig-zag (you are making an

up-and-down curve that transects the vertical center of the membrane). This will allow the pieces of the membrane to be put back together exactly for analysis of the Southern blot scan, and allows the hybridization to be done in a small tray with one half over the other, using less hybridization solution. Up to two membranes, four half-membranes, can be placed in one tray for hybridization at a time.

16. The flanking plasmid sequences from the insert PCR-probe are about 180 bp. These plasmid sequences do not cross-hybridize with eukaryotic DNA or lambda DNA, but may cross-hybridize with other MWMs.
17. The MWM probe can be stored at –20 °C. However, the isotope ^{32}P has a half-life of 14 days. Thus, when the stored MWM probe was made using a ^{32}P batch that was 2 weeks older than the freshly made specific probe, use 2 μL MWM probe (instead of 1 μL). For a difference of 4 weeks, 4 μL of MWM probe is used, and so on (with a maximum of 2 months).
18. Alternatively the labeled probe and non-incorporated dNTPs can be separated by using a Sephadex column. Omitting this step does not influence your hybridization result.
19. For some probes more stringent washing is desired due to a-specific hybridization of the probe to slightly homologous regions in the genome. For a GC-rich probe that recognizes a repetitive region, the specific signal can be much higher than 30 cps even after washing with wash buffer 4. In these cases more stringent hybridization condition are required [27].

Acknowledgement

Patrick van der Vliet and Silvère van der Maarel for critical reading.

References

1. Hewitt JE, Lyle R, Clark LN et al (1994) Analysis of the tandem repeat locus D4Z4 associated with facioscapulohumeral muscular dystrophy. Hum Mol Genet 3:1287–1295
2. Snider L, Asawachaicharn A, Tyler AE et al (2009) RNA transcripts, miRNA-sized fragments and proteins produced from D4Z4 units: new candidates for the pathophysiology of facioscapulohumeral dystrophy. Hum Mol Genet 18:2414–2430
3. Dixit M, Ansseau E, Tassin A et al (2007) DUX4, a candidate gene of facioscapulohumeral muscular dystrophy, encodes a transcriptional activator of PITX1. Proc Natl Acad Sci U S A 104:18157–18162
4. Lemmers RJLF, van der Vliet PJ, Klooster R et al (2010) A unifying genetic model for facioscapulohumeral muscular dystrophy. Science 329:1650–1653
5. Lunt PW, Jardine PE, Koch MC et al (1995) Correlation between fragment size at D4F104S1 and age at onset or at wheelchair use, with a possible generational effect, accounts for much phenotypic variation in 4q35- facioscapulohumeral muscular dystrophy (FSHD). Hum Mol Genet 4:951–958

6. van Deutekom JC, Wijmenga C, van Tienhoven EA et al (1993) FSHD associated DNA rearrangements are due to deletions of integral copies of a 3.2 kb tandemly repeated unit. Hum Mol Genet 2:2037–2042
7. Wijmenga C, Hewitt JE, Sandkuijl LA et al (1992) Chromosome 4q DNA rearrangements associated with facioscapulohumeral muscular dystrophy. Nat Genet 2:26–30
8. Snider L, Geng LN, Lemmers RJLF et al (2010) Facioscapulohumeral dystrophy: incomplete suppression of a retrotransposed gene. PLoS Genet 6, e1001181
9. Tassin A, Laoudj-Chenivesse D, Vanderplanck C et al (2013) DUX4 expression in FSHD muscle cells: how could such a rare protein cause a myopathy? J Cell Mol Med 17:76–89
10. Lemmers RJLF, Goeman JJ, van der Vliet PJ et al (2015) Inter-individual differences in CpG methylation at D4Z4 correlate with clinical variability in FSHD1 and FSHD2. Hum Mol Genet 24:659–669
11. Lemmers RJLF, Tawil R, Petek LM et al (2012) Digenic inheritance of an SMCHD1 mutation and an FSHD-permissive D4Z4 allele causes facioscapulohumeral muscular dystrophy type 2. Nat Genet 44:1370–1374
12. Lemmers RJLF, de Kievit P, Sandkuijl L et al (2002) Facioscapulohumeral muscular dystrophy is uniquely associated with one of the two variants of the 4q subtelomere. Nat Genet 32:235–236
13. Bakker E, Wijmenga C, Vossen RH et al (1995) The FSHD-linked locus D4F104S1 (p13E-11) on 4q35 has a homologue on 10qter. Muscle Nerve 2:39–44
14. van Deutekom JC, Bakker E, Lemmers RJLF et al (1996) Evidence for subtelomeric exchange of 3.3 kb tandemly repeated units between chromosomes 4q35 and 10q26: implications for genetic counselling and etiology of FSHD1. Hum Mol Genet 5:1997–2003
15. Lemmers RJLF, van der Vliet PJ, van der Gaag KJ et al (2010) Worldwide population analysis of the 4q and 10q subtelomeres identifies only four discrete duplication events in human evolution. Am J Hum Genet 86:364–377
16. Upadhyaya M, Maynard J, Osborn M et al (1995) Germinal mosaicism in facioscapulohumeral muscular dystrophy (FSHD). Muscle Nerve 2:45–9
17. van der Maarel SM, Deidda G, Lemmers RJLF et al (2000) De novo facioscapulohumeral muscular dystrophy: frequent somatic mosaicism, sex-dependent phenotype, and the role of mitotic transchromosomal repeat interaction between chromosomes 4 and 10. Am J Hum Genet 66:26–35
18. Deidda G, Cacurri S, Piazzo N et al (1996) Direct detection of 4q35 rearrangements implicated in facioscapulohumeral muscular dystrophy (FSHD). J Med Genet 33:361–365
19. Lemmers RJLF, de Kievit P, van Geel M et al (2001) Complete allele information in the diagnosis of facioscapulohumeral muscular dystrophy by triple DNA analysis. Ann Neurol 50:816–819
20. Wijmenga C, van Deutekom JC, Hewitt JE et al (1994) Pulsed-field gel electrophoresis of the D4F104S1 locus reveals the size and the parental origin of the facioscapulohumeral muscular dystrophy (FSHD)-associated deletions. Genomics 19:21–26
21. Nguyen K, Walrafen P, Bernard R et al (2011) Molecular combing reveals allelic combinations in facioscapulohumeral dystrophy. Ann Neurol 70:627–633
22. den Dunnen JT, van Ommen GJ (1993) Methods for pulsed-field gel electrophoresis. Appl Biochem Biotechnol 38:161–177
23. Lemmers RJLF, Wohlgemuth M, van der Gaag KJ et al (2007) Specific sequence variations within the 4q35 region are associated with facioscapulohumeral muscular dystrophy. Am J Hum Genet 81:884–894
24. Calandra P, Cascino I, Lemmers RJLF et al. (2016) Allele-specific DNA hypomethylation characterises FSHD1 and FSHD2. J Med Genet. [Epub ahead of print]
25. Hartweck LM, Anderson LJ, Lemmers RJLF et al (2013) A focal domain of extreme demethylation within D4Z4 in FSHD2. Neurology 80:392–399
26. Jones TI, Yan C, Sapp PC et al (2014) Identifying diagnostic DNA methylation profiles for facioscapulohumeral muscular dystrophy in blood and saliva using bisulfite sequencing. Clin Epigenetics 6:23
27. Ehrlich M1, Jackson K, Tsumagari K, et al (2007) Hybridization analysis of D4Z4 repeat arrays linked to FSHD. Chromosoma 116:107–116

Chapter 8

Analysis of Copy Number Variation Using the Paralogue Ratio Test (PRT)

Edward J. Hollox

Abstract

Copy number variation (CNV), where a segment of DNA differs in copy number between different individuals, is an extensive and often underappreciated source of genetic variation within species. However, reliably determining copy number of a particular DNA sequence for a large number of samples can be challenging. Here, I describe and review the paralogue ratio test (PRT) in detail. PRT was developed to robustly type the CNV of the beta-defensin locus using small amounts of genomic DNA in a high-throughput manner, and has been applied successfully at many other loci. I discuss the strategies for designing successful PRT assays using both manual and bioinformatics methods, how to optimize experimental conditions, and approaches for analyzing the data. I discuss strengths and weaknesses of the approach, and how to troubleshoot results, as well as the range of problems to which PRT can be a potential solution.

Key words Copy number variation, CNV, PRT, PCR, Deletion, Duplication, Beta-defensin, Genotyping, High-throughput

1 Introduction

Copy number variation (CNV), where a segment of DNA differs in copy number between different individuals, is an extensive and often underappreciated source of genetic variation within species [1]. It encompasses deletions and duplications as well as more complex multiallelic CNV (mCNV), where there may be many different copy number alleles within a population. An important difference between mCNVs and most (but not all) deletions and duplications is that mCNVs may have a considerably higher mutation than most deletions and duplications, and certainly compared to nucleotide substitutions [2]. This means that new mCNV alleles can be generated by recurrent mutation, and may not show strong linkage disequilibrium with neighboring single-nucleotide polymorphisms (SNPs) [3, 4].

In humans, there has been much interest in the relationship between CNV and disease [5–8]. Although the effect of rare deletions and duplications in disease is now well established [9],

Stefan J. White and Stuart Cantsilieris (eds.), *Genotyping: Methods and Protocols*, Methods in Molecular Biology, vol. 1492, DOI 10.1007/978-1-4939-6442-0_8, the role of common CNV is not yet clear. There are some examples of common deletions being risk alleles for various common diseases, such as deletion of the late cornified envelope genes LCE3B and LCE3C as a risk allele for psoriasis [10], and a deletion allele upstream of IRGM as a Crohn's disease risk allele [11]. These can be identified by SNP-based genome-wide association studies, because some SNPs will be in linkage disequilibrium with a simple diallelic deletion which has occurred once in evolutionary history and is not a product of recurrent mutation. For duplications and mCNVs, which have been generated by recurrent mutation, in most cases different copy number alleles are not in LD with neighboring SNPs and are effectively invisible to genome-wide SNP association studies, although it should be noted that some duplications and mCNVs can be effectively imputed from huge, dense SNP genotype data [3]. There is currently no robust, high-throughput, cost-effective method for genome-wide analysis of CNVs on thousands of samples, which has limited analysis of CNV to locus-specific studies. While there have been some successes, in general these studies have been limited by sample sizes powered to detect only very strong genetic effects, which are unlikely in common disease, and have used error-prone methods, further limiting the power of the data with noise [12–14].

Reliably determining copy number of a particular DNA sequence for a large number of samples can be challenging. The paralogue ratio test (PRT) is a particular form of quantitative PCR that can be used to type the copy number of a particular locus of thousands of samples using typically 10–20 ng of genomic DNA [15]. Many comparative studies and reviews have found it considerably more accurate and precise than real-time quantitative PCR [12, 16–19]. It is cost effective for large-scale studies since it uses equipment typically found in the molecular genetics laboratory, with the most significant cost being the requirement for a capillary electrophoresis machine such as the Applied Biosystems 3130xl and its associated consumables. While it can be used to robustly type simple deletions and duplications, it was developed particularly to type copy numbers of multiallelic copy number variations (mCNVs), and can be applied to mCNVs where copy numbers range between 0 and 10. In theory, higher copy numbers can be determined, but the method becomes increasingly imprecise at higher copy numbers, although repeat testing of the sample can mitigate this effect.

1.1 Development of the Paralogue Ratio Test

The PRT was developed to robustly type the mCNV of the beta-defensin locus using small amounts of genomic DNA in a high-throughput manner [15]. When it was developed, it was clear that the approach could be applied to other mCNVs in the genome, and indeed a very similar approach had previously been applied to type aneuploidies [20]. It is also similar to the competitive PCR approach that has been used to measure DNA

Table 1
Published examples of PRT assays

Gene or region	Publication	Comments
Beta-defensin cluster	[15, 17]	Initially a single PRT, then redesigned as a triplex PRT
Low-affinity Fc gamma receptor cluster	[23]	
Complement C4	[24]	
Haptoglobin-related protein *HPR*	[25]	
Alpha-defensin 1 (*DEFA1A3*)	[26]	
Salivary agglutinin (*DMBT1*)	[27]	Multiple PRTs for two CNVs
CCL3L1/CCL4L1	[28, 29]	A small improvement in the protocol in the more recent paper
Chromosome 13 and chromosome 18	[30]	Used for trisomy 18 and trisomy 13 detection
SLC2A3	[31]	
22q11.2 deletion	[32]	
Amylase region (*AMY1* and *AMY2*)	[33]	
Chromosome Y palindrome arm	[34]	
SIRPB1	[35]	

dosage and cDNA levels of a particular gene [21, 22]. PRT has used for many of the well-established mCNVs in the human genome, where associations with disease have been suggested or found (Table 1). In general, application of a reliable CNV assay in a well powered cohort has shown no evidence of disease association, contrary to previous results, typically gathered using real-time qPCR. However, in some cases an association with a clinical phenotype has been found, and in perhaps still the most robust finding, an association between beta-defensin copy number and psoriasis risk has been replicated [36, 37].

In this chapter I will describe the principles of PRT and approaches to developing new PRT assays. My laboratory has generally found developing new PRT assays to have a steep learning curve and to be sometimes frustrating, with several rounds of optimization and assay re-design required. The aim of this chapter is to help climb the steep learning curve, with the intention of developing a robust high-throughput cost-effective CNV assay ready to test hundreds of genomic DNA samples.

1.2 Molecular Principle of PRTs

PRT can be regarded as a form of quantitative PCR. In conventional quantitative PCR (qPCR), two pairs of distinct PCR primers are designed to amplify a reference locus, which is assumed not to vary between samples, and a test locus corresponding to the gene that is variable between samples. Comparison between the amplification of the test and reference locus, usually by real-time monitoring of amplification products, can allow quantification of the reference locus compared to the test locus. The key difference between conventional qPCR and PRT is that in PRT a single pair of primers is designed to amplify both the test and reference loci. Then, following PCR with one of the pair of primers fluorescently labeled, the reference and test amplicons are separated by capillary electrophoresis and the amplicons calculated by measuring the area of each peak (Fig. 1).

Capillary electrophoresis of test and reference amplicons separated by size is the most commonly implemented method of measuring the relative amounts of the two amplicons, but other methods have been used. Amplification using non-fluorescently labeled primers followed by separation on ethidium bromide-stained agarose gels and densitometry of gel image has been used successfully as a low-tech low-cost alternative, but, given the use of an intercalating dye and limited resolution of such an approach, this is probably limited to the lower range of copy numbers. Furthermore, the difference in size between the test and reference amplicons has to be sufficient to allow clear resolution on an agarose gel. If a single-nucleotide change, instead of a small deletion, is used to distinguish the test amplicon from reference amplicon, then any method that can quantify such a single-nucleotide difference can be used to measure the relative amounts of test and reference. Taqman hydrolysis probes in real-time qPCR have been used [30] (Fig. 2a), as has pyrosequencing [18, 38] (Fig. 2b), mass spectrometry [35], and, as in the original PRT publication, restriction enzyme digestion [15].

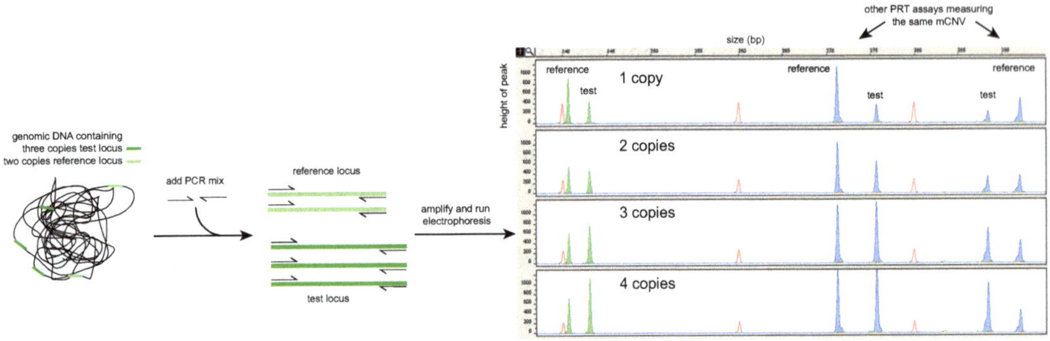

Fig. 1 The principle of PRT for measuring copy number. From *left* to *right*, this figure shows a single PRT analysis of a genomic DNA sample that has three copies of the test locus. Following addition of a single pair of primers and PCR amplification, the reference and test products are separated on the basis of size by capillary electrophoresis, as shown by an electropherogram on the right-hand side of the figure. Also shown in *blue* are two other PRT assays that can be electrophoresed on the same capillary

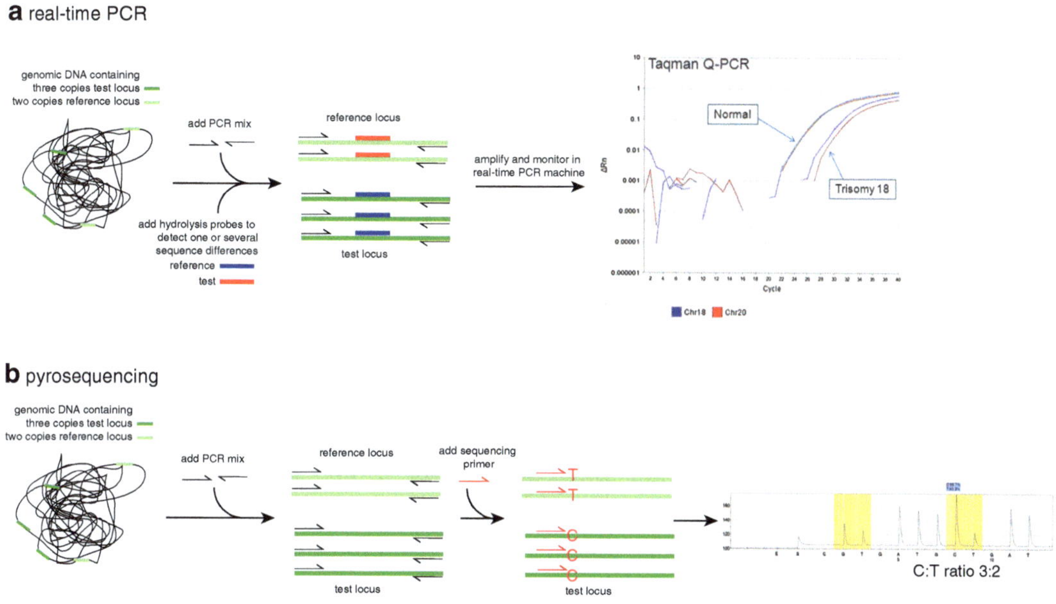

Fig. 2 Alternative approaches for distinguishing test and reference amplicons in PRT. (**a**) Using real-time PCR. As in Fig. 1, a three copy test locus is amplified. The reference and test amplicons can be distinguished by one or more single-nucleotide differences, so that hybridization with a sequence-specific hydrolysis probe (such as a Taqman probe) can be used to continuously monitor the two amplication products during PCR. In this example, the reference locus is on chromosome 20 and the test is on chromosome 18, and the amplification plot shows results from a normal genomic DNA and genomic DNA from an individual with trisomy 18 [30]. (**b**) Using Pyrosequencing. As in Fig. 1, a three copy test locus is amplified. The reference and test amplicons can be distinguished by one or more single-nucleotide differences, so a sequencing primer (*red arrow*) followed by Pyrosequencing can be used to distinguish relative amounts of reference (T variant) and test (C variant) amplicons [18]

The advantage of PRT over other qPCR methods appears to be that because the same pair of primers anneal to test and reference loci, and the two amplicons are often very similar in sequence, the kinetics of amplification of the two amplicons are very similar. This can be seen in PRT where the amplicon amplification is followed by real-time PCR (Fig. 2a [30]). The result of this is that quantification of the amplicon amounts at the PCR endpoint is an accurate representation of the relative amounts of starting target sequences.

2 Materials

One of the principal advantages of the PRT method is that the reagents and equipment required are generally available in most molecular genetic labs, since it is essentially a particular form of PCR. We prepare all PCR reagents under PCR clean conditions, and aliquot to minimize freeze-thawing, which we find particularly important for fluorescently labeled primers. We routinely use ABI

Veriti thermal cyclers (Life Technologies, Thermo Fisher Scientific), but have used other thermal cyclers as well with success. The one piece of equipment which may be less accessible to an individual, but is often available through a shared genomics service, is an Applied Biosystems (Life Technologies, Thermo Fisher Scientific) capillary electrophoresis machine. We routinely use a 16-capillary ABI3130xl, but an 8- or 96-capillary machine will be as effective. Use of this equipment is as the manufacturer's guidelines, with electrophoresis buffers made using HPLC-grade water (Fisher Scientific, Loughborough, UK).

2.1 PCR Reagents

10× LD (low dNTP) buffer: 500 mM Tris–HCl pH 8.8, 125 mM $(NH_4)_2SO_4$, 14 mM $MgCl_2$, 75 mM 2-mercaptoethanol (reagent grade, Fisher Scientific, Loughborough, UK), 2 mM dATP (sodium salt, Promega), 2 mM dCTP (sodium salt), 2 mM dGTP (sodium salt), 2 mM dTTP (sodium salt), 1.25 mg/ml unacetylated bovine serum albumin (Ambion Inc).

10 μM Forward primer (reverse phase purified, 5′ labeled with a fluorescent dye detected by the capillary electrophoresis machine).

10 μM Reverse primer (reverse phase purified).

Taq DNA polymerase (5 units/μl).

Molecular biology grade H_2O.

Genomic DNA (5–10 ng/μl) (*see* **Notes 1** and **2**).

2.2 Variation in PCR Components

The buffer we use (10× LD buffer) is one that we make in-house. We find that this performs well on its own (e.g., [17, 39]), but have also found that combining this with commercial ammonium sulfate-based 10× PCR buffer (containing 15 mM $MgCl_2$, routinely supplied with *Taq* DNA polymerase) at a final conc of 1× can increase consistency of signal across samples, probably due to the increased amount of free Mg^{2+} ions and/or increased buffering capability of the PCR reaction [27]. Although we routinely use Kapa *Taq* (KAPA Biosystems), we have used routine *Taq* DNA polymerase from several other manufacturers (Invitrogen, Bioline, ThermoFisher) with success, so a special *Taq* DNA polymerase is not required. Annealing temperature of the PCR and total number of PCR cycles should be determined empirically to give clear expected peaks with peak areas, following capillary electrophoresis, between 400 and 40,000.

3 Methods

3.1 Approaches to Designing a CNV Assay Using PRT

1. CNV can involve a whole gene, part of a gene, or no gene at all, and knowing the extent of a CNV region not only suggests what the phenotypic consequences of that CNV may be, if any, but it

also provides the range over which PRT assays that measure the same CNV may be designed. Sensible guesses about the extent of CNV can often be made from genomic annotations, such as the extent of segmental duplications . However, the best evidence for CNV extent is from genome-wide CNV analyses using either array comparative genomic hybridization or next-generation sequencing sequence read depth approaches. Evidence for the extent of particular region of CNV, even from a small number of samples, is very valuable in refining the region where PRT assays need to be developed (*see* **Note 3**).

2. PRTs can be divided into two types, depending on the location of the reference target compared to the test target. Trans-PRTs have the reference amplicon on a different chromosome, or greater than 500 kb away from the test amplicon, and cis-PRTs have the test and reference amplicons closer together. In both cases, there needs to be some evidence that the reference amplicon itself does not map to a copy number variable region (from the Database of Genomic Variants, for example). For trans PRTs, the increased distance between reference and test regions means that there is less likelihood of genome rearrangements affecting both regions, and therefore that the reference region is less likely to be CNV. For cis-PRTs, the proximity of test and reference regions means that genome rearrangements affecting both regions are more likely, and indeed, if both are on a larger segmental duplication, may be part of the same CNV. Therefore extra care must be taken in selecting an appropriate non-CNV reference amplicon. In practice, we have found that cis-PRTs generally require less optimization and perform better when compared to trans-PRTs, and that an ideal strategy for measuring CNV of a region is to have at least one trans-PRT in combination with several cis-PRTs. This allows results from multiple assays to be compared against each other, verifying that all are measuring the same CNV, and therefore combining data from several assays increasing accuracy and precision of the final copy number call [28].

3. There are several approaches to identifying potential PRT primers. The most straightforward is to use the software PRTprimer [40]. This searches a reference genome for all potential PRT primers, within certain parameters such as distance between test and reference region, size of amplicons, and primer design variables. It can allow for multiple copies of the test locus in the assembly, and screens primers for any overlap with nucleotides that show known single-nucleotide variation. PRT candidates for the human genome identified by PRTprimer are publicly available as a searchable database online (prtprimer.org). PRTprimer has been used on the mouse and rhesus macaque genome, and can be freely downloaded and run on any reference genome; *see* **Note 4**.

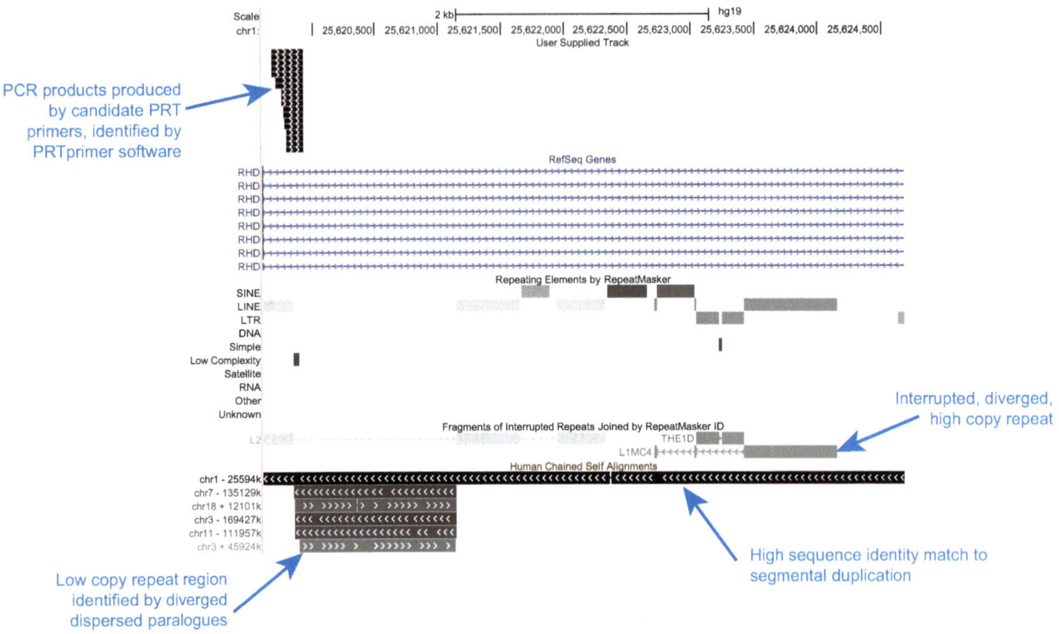

Fig. 3 Annotation tracks on the UCSC Human Genome Browser useful for PRT design. Part of an intron of the *RHD* gene highlighting different annotation tracks that can aid in designing new PRT assays. The "User Supplied Track" is the track supplied by the software tool PRTprimer. Annotations provided by RepeatMasker and the human self-alignment are also shown

4. There are other ways of designing PRT primers based on manual inspection of a reference genome (Fig. 3). For the human genome, there are several annotations available on the UCSC Genome Browser that are helpful for this task [41]. Perhaps the most useful is the self-chain track, which represents the human genome aligned to itself, and reflects regions of duplication and other sequence similarities. Other useful annotations include segmental duplications [42] and interrupted repeats, generated by Repeatmasker [43]. Careful design of primers that anneal to diverged repeats can provide useful candidate PRTs, and indeed this was the approach taken in identifying candidate PRTs in the paper introducing PRT [15].

5. Selection of effective PRTs from several candidate PRTs is an empirical process. We have found that identification of positive controls of known copy number is essential to verify the reliability of a candidate PRT assay. For the human genome, this is straightforward, as there is a set of shared samples which are publicly available: the HapMap samples and, subsequently, 1000 Genomes sample. For these samples, there are data on genome-wide copy number variation (arrayCGH and short read sequence read depth), allowing selection of 6–7 samples of known copy number samples [3, 44–47]. For mCNVs that are not at polymorphic frequencies, or for other species where

a set of diverse DNA samples are not publicly available or do not have extensive genome-wide data, selection of positive controls is more difficult. Nevertheless, copy number estimates from a small number of positive control DNA samples generated by short-read sequence read depth, MLPA, MAPH, or fiber-FISH, are essential to validate candidate PRTs and to act as positive controls during routine use of PRT assays.

6. PRT optimization follows essentially the same path as optimization of any other PCR. There are two aims, firstly to achieve clear specific amplification of test and reference amplicons, and secondly to show that the ratio of test:reference covaries with copy number. Although the first stage can be shown with any appropriate genomic DNA of the relevant species, the second requires positive controls of known copy number.

7. When several PRTs have been validated (typically 3–4) it is often convenient to multiplex them in preparation for high-throughput typing of mCNV. The multiplex can occur at two stages—at the PCR stage or at the electrophoresis stage [17, 23, 28] (Fig. 4). If the selected PRTs can all perform well under the same PCR conditions, and all the amplicons are sufficiently distant from each other to limit the production of larger amplicons from the same PCR primers, then all PCR primers can be combined in one PCR reaction in a multiplex PCR . This has the advantage of reducing consumable cost, sample mixup error, and assay time, compared to doing three or four separate PCRs. If the selected PRTs have different PCR optimization conditions, then they need to be amplified in separate reactions

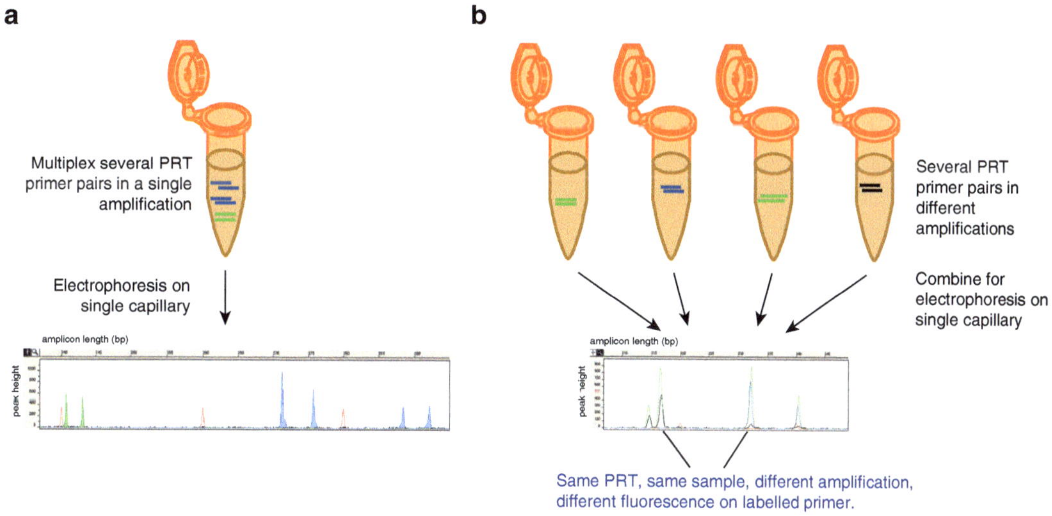

Fig. 4 Strategies for multiplexing PRTs. PRTs can be multiplexed at the amplification stage (**a**) or at the electrophoresis stage (**b**) to minimize consumable costs

but can be combined at the electrophoresis stage, so that multiple PRT products can be run on a single capillary. This still gives cost savings, and has the advantage that a single PRT can be in two separate reactions and run on the same capillary, by using differently fluorescently labeled primers. This can increase precision of a particular PRT.

3.2 Typical Setup of PRT

1. Set up a PCR under PCR-clean conditions (laminar flow hood, dedicated pipettes, and plastics, clear of PCR amplified products), in suitable PCR tube or PCR plate. To 6.9 μl molecular biology-grade water, add 1 μl 10 × LD buffer, 0.5 μl 10 μM primer F, 0.5 μl 10 μM primer R, 0.1 μl *Taq* DNA polymerase, and 1 μl of genomic DNA at 3–10 ng/μl. Include a negative control in every PCR reaction, which does not include genomic DNA, to check for contamination of solutions by PCR product or genomic DNA.
2. We routinely run PRT PCRs in 96-well plates, with the last row of 8-wells occupied by six positive control reactions and two negative controls (PCR without genomic DNA). The six positive controls are of different copy number and ensure that the PRT has worked correctly and help correct for batch effects in the subsequent analysis stages. Nevertheless, we would still recommend distributing cases and controls randomly across all plates, if possible, to minimize the likelihood of batch effects in a case–control study, for example. This setup allows amplification of 88 test samples, although of course fewer samples can be run, occupying fewer wells in the plate.
3. Cycle the PCR reactions in a thermal cycler, according to the conditions given in Table 2.
4. At the end of cycling, add 0.5–1.0 μl of each PRT to 10 μl of a 1:100 mix of Mapmarker ROX-400 (Eurogentec, Fawley, UK) and deionized formamide. Following denaturation at 95 °C for 3 min and snap-cooling on ice, load the sample plate on the ABI and electrophorese according to standard conditions recommended by the manufacturer.

Table 2
PCR conditions for routine PRTs

Step	Stage	Temperature (°C)	Time (s)
1	Initial denaturation	95	120
2	Denaturation	95	30
3	Annealing	58	30
4	Extension	72	30
5	Go to stage 2, 24 times		
6	Final extension	72	30 min

3.3 Analysis of PRT Data

1. Following capillary electrophoresis, the area under each peak is called using Genemapper or the free software PeakScanner (http://www.appliedbiosystems.com). Our first-quality control threshold is the area under the peak, with PRTs showing a peak area value less than 400 or greater than 40000 being discarded. Samples with weak but visible peaks can be rerun with either more of the PCR product (up to a maximum of around 2 μl) or with a longer electrophoresis injection time. Samples with peaks that are over the threshold can be diluted and run again.
2. The test value is then divided by the reference value to give a number that is known as the unnormalized raw PRT value, directly reflecting copy number at the test locus. The unnormalized values for the positive controls and the matched known copy numbers are then compared in a linear regression (Fig. 5). Following manual inspection of the regression line, the regression equation is then used to normalize each raw PRT value across the experiment. This is repeated for each PRT in the multiplexed experiment. Normalized raw PRT values for each sample are then combined into a single value for each sample, usually by using the mean or sometimes by the value of the first principal component. Samples that initially failed the peak area QC threshold, and have been rerun on the dilution following dilution or increased amount/injection time, can be normalized using the same regression equation as the original electrophoresis run.
3. The next stages are critically important quality control check of your data, and should be always done on large datasets (>100 samples). Firstly, a histogram of the combined normalized raw PRT value for each sample should be plotted. This should show clear distinct peaks reflecting real integer copy numbers rather than a broad spread of results. The peaks should be particularly clear at the low copy number ranges, and may begin

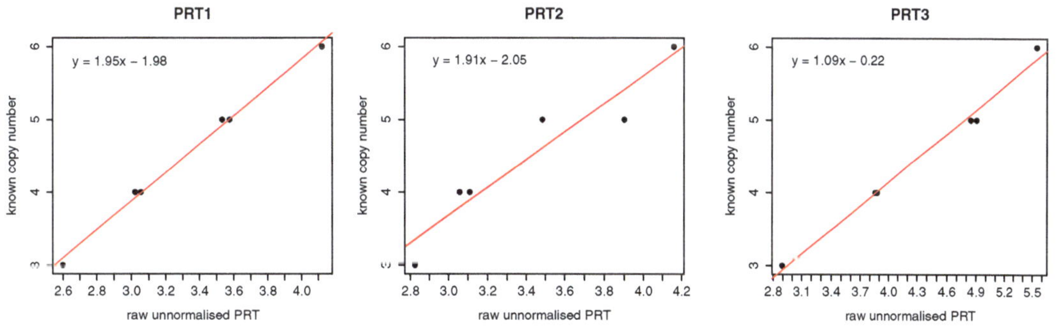

Fig. 5 Normalization plots for three example PRTs. Three scatterplots are shown, with each point representing a result from a positive control sample from a particular PRT experiment. In each case, the six positive controls have a known copy number (*y* axis) and a raw unnormalized PRT value (*x*-axis). The *red line* shows the linear regression for the data, and the regression equation is also shown at the top left of each plot

to merge at higher copy numbers (Fig. 6). Secondly, scatterplots comparing the normalized raw ratios of each individual PRT should show some evidence clustering (Fig. 6c), and these plots stratified by batch or experiment help to identify batch-to-batch variation. Optionally, a censoring quality control step can be introduced, rejecting samples where the coefficient of variation of the normalized raw ratios is more than a given threshold (we have used 0.15 as that threshold), with the expectation that almost all of the samples will pass this quality control step and only a few fail and need to be retested. Once these checks have been done, your data will now be a list of samples with a corresponding mean normalized raw PRT value reflecting copy number of the locus of interest.

3.4 Calling Integer Copy Number

1. The data can either be analyzed further as raw copy number estimates, or integer diploid copy numbers [2–4] called from the data (*see* **Note 5**). There are three approaches to calling integer copy number, described below, and shown in Fig. 6.
2. Using raw copy number:

   The principal argument for using the mean normalized raw PRT value (raw copy number) is that the error in the method is incorporated into subsequent analysis. This is important for association analysis, where calling integer copy numbers and then using those to test for association with a trait may lead to false-positive results. Use of raw copy number has been used for testing association with both quantitative traits and case–control cohorts. Importantly, the single raw value can be easily incorporated into more complex statistical tests, incorporating for example co-factors and covariates [48]. However, raw copy numbers do not make much sense biologically (a copy number of 1.9 and 2.1 will reflect a real copy number of 2) and for some analyses, such as variant frequency calculation or confirmation of inheritance in pedigrees, analysis of integer copy number is more sensible.
3. Calling integer copy number by binning:

   This involves simply providing thresholds (often arbitrary) which can then be used to bin samples into integer copy number classes based on their raw copy number. If the raw copy number data clearly partition into different clusters with no overlap or merging between clusters, then simple binning is a perfectly sensible approach. If, however, there is some degree of overlap between clusters, then binning into integer copy number classes will convert copy number calls with error into apparently error-free copy number calls. While suitable for some purposes, this may lead to false-positive genetic associations with traits, for example.

Fig. 6 (continued) The Gaussian curves are shown superimposed on the left-hand histogram. (**c**) Calling copy number using a maximum-likelihood approach. The *left panel* shows a scatterplot matrix of three PRTs run on the same set of samples

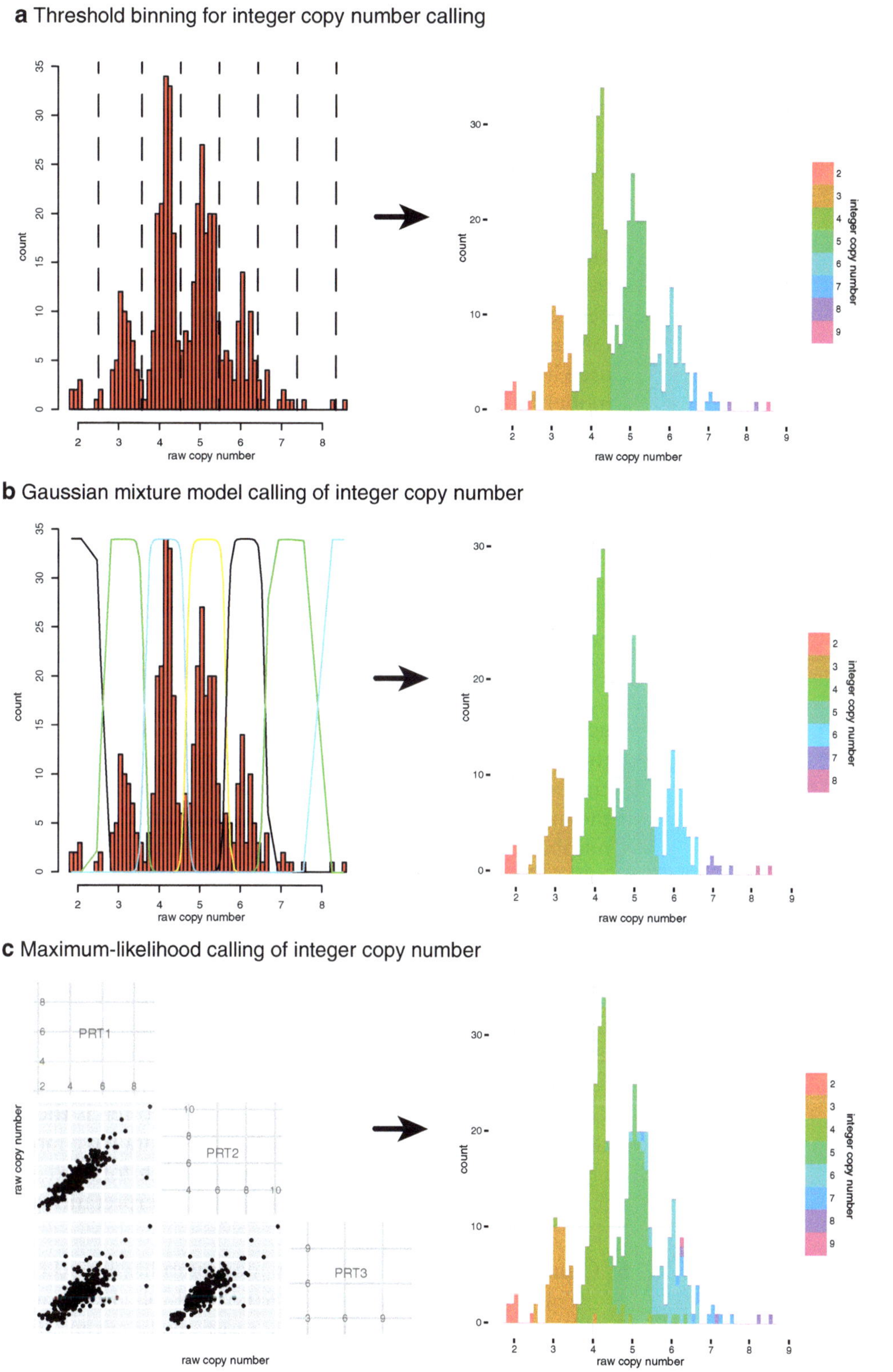

Fig. 6 Calling integer copy number from raw normalized PRT values. In each part of the figure, the left histogram shows the distribution of raw normalized PRT results (*x*-axis, raw copy number) and the right histogram shows the same data colored according to the final integer copy number called. Note that the colors of 7, 8, and 9 copies differ between histograms. (**a**) Using threshold bins to call copy number, with each threshold set halfway between each integer. (**b**) Gaussian mixture modeling of integer copy number.

4. Calling integer copy number by Gaussian mixture modeling (GMM):

   This is essentially a more sophisticated approach to copy number binning. Instead of an arbitrary threshold used to bin raw copy number, several Gaussian curves are fitted to the data with each Gaussian curve reflecting the probability of a given integer copy number call. Therefore, a key difference between this approach and simple binning is that each integer copy number call is accompanied by a probability of that copy number call, and in this way the error of each call is recorded and can be used in subsequent downstream analyses. The Gaussian mixture model fitting approach can also be incorporated into tests of association with case–control status or a quantitative trait, by allowing particular parameters to vary and comparing the likelihood of a GMM with a variable parameter against a null model of no difference between the GMMs of cases and controls, for example. Integer copy number calling and association tests are implemented by the package CNVtools [49] in the statistical language R, but at present only simpler association models can be tested. Other R packages such as CNVassoc [50, 51] and CNVCALL [52] implement similar approaches.

   Overall, the GMM approach has the best of all worlds: integer copy number calling with associated measures of error, and the possibility of a robust statistical framework to test for association with traits. However, it does have limitations. Perhaps the most important limitation is the importance of choosing the number of Gaussian curves to fit to the data (in statistics this is called the number of components of the GMM). In data that clearly cluster in a histogram, the number of peaks can be seen by eye and this would be used to choose the number of Gaussian curves. In data that doesn't cluster quite as well, a number of different components can be chosen and the one that fits the data best, as measured by a statistical goodness-of-fit test such as the Bayesian Information Criterion, is chosen. CNVtools, for example, can implement this approach. However, where two GMMs fit the data equally well (or, as more often is the case in these circumstances, equally badly), or the data have rare outliers, fitting an appropriate GMM is not straightforward and different GMMs can lead to very different interpretations of the data (*see* **Note 6**).

5. Calling integer copy number by maximum likelihood (ML):

   This approach takes results from multiple PRTs and asks which integer copy number is most likely given the data observed, and is most effective when each PRT assay is run in duplicate, or more times. It calls integer copy number by assuming a Gaussian distribution with the mean being the average normalized raw PRT value across the repeated measurements of the same sample, and the standard deviation being the observed standard deviation of the repeated measurements. This Gaussian distribution is then used to calculate the relative likelihood of the data reflecting each integer number, usually between 0 and 10. These likelihoods can then be combined with likelihoods from other assays to give a likelihood of each integer copy number, with the integer copy number with the largest likelihood chosen as correct [17].

This approach has the advantage, like the GMM, that each sample copy number call is accompanied by an error value reflecting the confidence in each copy number call. It is distinct from the GMM because it provides the copy number call and error value given the results from one sample, rather than requiring a large number of samples to fit a GMM. This has the advantage that estimates of integer copy number are not dependent on the sample size and that the estimate of error is sample specific, but has the disadvantage that the extra information provided by large number of samples cannot be used to call integer copy number. The ML approach is also sensitive to unusually small standard deviations in a particular Gaussian curve, sometimes generated by chance. The ML approach can be used on non-duplicate measurements, with the SD for each assay estimated from repeat testing of a control, but this loses the advantage of the error rate being specific to that particular sample. We have also found the ML approach to be unstable in some situations, particularly at higher copy numbers.

4 Notes

1. PRT and DNA quality:

   As well as the parameters that affect PRT assays discussed earlier in this chapter, there are other, more nebulous, factors that affect the performance of PRT assays, and these factors can vary in importance from assay to assay. For example, we have found that some, but not all, PRTs are vulnerable to differences between DNA cohorts. The reasons for this are unclear, and there may not be one single reason, but in one investigation the existence of thermodynamically ultra-fastened (TUF) regions in the genome affected the relative efficiencies of amplification in PRTs [53]. The importance of TUF regions in restricting amplification from genomic DNA depends on the physical structure of the genomic DNA, and highly fragmented/sheared genomic DNA is more resistant to this process. This leads to the prediction that "high-quality" genomic DNA, consisting of large DNA molecules, will yield less reliable results from some PRT assays where one of the amplicons is within or nearby a TUF region. Indeed we have found that highly sheared DNA performs perfectly well in PRT assays. There are two practical approaches to minimizing this effect in problematic assays. Firstly, introducing an initial denaturation step of 98° for 3 min at the start of the PCR can improve some PRTs. Secondly, shearing of genomic DNA by sonication or by restriction enzyme treatment may prove useful. Also, the positive controls must be of similar quality to the DNA cohort being tested—using positive control DNAs comprised of large high-molecular-weight fragments together with a DNA cohort consisting of sheared DNA may affect results. Analysis of representative DNA samples on a 0.8 % agarose gel or an Agilent Bioanalyzer is sufficient to determine the structural integrity of the genomic DNA.

2. PRT and DNA source:

   In contrast to the influence of DNA quality, we have found that DNA source is not important, and PRT has generated successful results from genomic DNA extracted from cells, saliva, mouthwash, vaginal swabs, fresh peripheral blood, and dried blood spots. Indeed, PRT has been successfully used on very small degraded fragments extracted from formalin-fixed paraffin-embedded material [30]. Use on FFPE material opens up a wide range of applications for PRT, for example on archive samples and on different tissues.

3. Refining CNV regions using PRTs:

   Defining the exact boundaries of CNV regions can be challenging but is important to establish the nature of any effect of the CNV; for example, whether a whole or part of a gene is within the CNV. One approach to help refine CNV boundaries is to design several PRTs spanning the region across the likely boundary site and compare copy number results for each separate PRT assay. This approach has limited resolution, because PRTs cannot be designed at regular intervals, for example, and because one region of CNV may be embedded within a region that shows an alternative pattern of CNV.

   An alternative approach is to use a set of samples where matching PRT data and genome-wide data (such as dense array CGH or next-generation sequence read depth) are available on a number of samples. The pairwise correlation coefficients between each array CGH probe (for example) and the copy number (determined by PRT) can be calculated, and the value plotted at the position of each probe across the region of interest. The CNV region is then shown as an area of high average correlation coefficient value, while the CNV boundaries are shown by a drop to a correlation coefficient of around zero. Such an approach was used on the human beta-defensin region, which is a mCNV embedded within a complex duplication rich region [54].

4. PRT and single-nucleotide variation:

   It has been noted by others that PRT, because it relies on PCR, can be vulnerable to single-nucleotide variation underneath the primers affecting annealing of the primers in an allele-specific manner [55]. The effect of this can be minimized by screening primers for known variation in a sequence variation database, selecting the lowest PCR annealing temperature that produces specific amplicons, and by incorporating more than one PRT to determine copy number. Certainly, in organisms such as human, with a deep catalogue of single-nucleotide variation, such problems are more easily avoided, but in organisms without such a database more care in PRT primer design may be required.

PRT should be used with caution if a well-assembled reference genome is not available. This is because PRT primer design relies on the test and reference amplicons being readily identified from the genome assembly, and that similar sequences are not present elsewhere in the genome but unassembled in the reference genome. Even in humans, unassembled sequences occur, and in poorly assembled complex genomes from other organisms PRT design in complex multiallelic loci may be constrained.

5. From diploid copy number to genotype:

It is worth remembering that PRT gives the total diploid copy number of a locus, which is the sum of the copy numbers across both homologous chromosomes, rather than a true genotype. This might not be important if any phenotypic effect is assumed to reflect gene dosage, but if a genotypic effect is suspected (for example—if a 3–1 genotype is different from a 2–2 genotype despite both having a copy number of 4) then knowing genotype information is important. It is also useful for phasing in copy number with surrounding SNP haplotypes for population genetic analysis.

Determining genotype from diploid copy number is difficult, and there are only two practical ways to do it. Firstly, by genotyping a large family and observing segregation patterns [27, 56]. Secondly, fitting the most likely frequency distribution of copy number alleles in a population that would combine to give the observed diploid copy number distribution, assuming Hardy-Weinberg equilibrium, and calculating individual probability of a particular genotype from that frequency distribution [57]. Both methods are less than ideal; yet an approach to determine copy number genotype molecularly in large numbers of samples is lacking.

6. How good is a Gaussian model for calling integer copy number?

Both MLE and GMM assume a Gaussian distribution of raw normalized PRT values. In theory, because these values are in fact a ratio of the test amplicon area and the reference amplicon area, we might expect the error distribution to reflect the ratio of two Gaussian distributions. This distribution, called the Cauchy distribution, is statistically horrendous, because the parameters of the distribution, such as the mean and variance, cannot be easily estimated from the data. The long tails characteristic of the Cauchy distribution can be mimicked by the similar t-distribution, and indeed CNVtools can fit a mixture of t-distributions to copy number data, in a manner analogous to fitting a mixture of Gaussian distributions. Thankfully, perhaps, empirical analysis of the error distribution of raw normalized PRT values shows a good fit with the Gaussian distribution [58], although the real data do show longer tails than a Gaussian distribution. These subtle differences in error distribution are very unlikely to have a significant effect on copy number calling.

Acknowledgements

I would like to thank all current and former members of the lab, but particularly Luciana Zuccherato, Robert Hardwick, and Adeolu Adewoye for providing example electropherograms for the figures, and Colin Veal for helpful comments on the manuscript.

References

1. Schrider DR, Hahn MW (2010) Gene copy-number polymorphism in nature. Proc Biol 277:3213–3221
2. Campbell CD, Eichler EE (2013) Properties and rates of germline mutations in humans. Trends Genet 29:575–584
3. Handsaker RE, Van Doren V, Berman JR et al (2015) Large multiallelic copy number variations in humans. Nat Genet 47:296–303
4. Locke DP, Sharp AJ, McCarroll SA et al (2006) Linkage disequilibrium and heritability of copy-number polymorphisms within duplicated regions of the human genome. Am J Hum Genet 79:275–290
5. Wain LV, Armour JAL, Tobin MD (2009) Genomic copy number variation, human health, and disease. Lancet 374:340–350
6. Zhang F, Gu W, Hurles ME et al (2009) Copy number variation in human health, disease, and evolution. Annu Rev Genom Hum G 10:451–481
7. Hollox EJ, Hoh B-P (2014) Human gene copy number variation and infectious disease. Hum Genet 133:1217–1233
8. Usher CL, McCarroll SA (2015) Complex and multi-allelic copy number variation in human disease. Brief Funct Genome 14:329–338
9. Iyer J, Girirajan S (2015) Gene discovery and functional assessment of rare copy-number variants in neurodevelopmental disorders. Brief Funct Genome 14:315–328
10. de Cid R, Riveira-Munoz E, Zeeuwen PL et al (2009) Deletion of the late cornified envelope LCE3B and LCE3C genes as a susceptibility factor for psoriasis. Nat Genet 41:211–215
11. McCarroll SA, Huett A, Kuballa P et al (2008) Deletion polymorphism upstream of IRGM associated with altered IRGM expression and Crohn's disease. Nat Genet 40:1107–1112
12. Cantsilieris S, Western PS, Baird PN et al (2014) Technical considerations for genotyping multi-allelic copy number variation (CNV), in regions of segmental duplication. BMC Genomics 15:329
13. Cantsilieris S, White SJ (2013) Correlating multiallelic copy number polymorphisms with disease susceptibility. Hum Mutat 34:1–13
14. Hollox EJ (2010) Beta-defensins and Crohn's disease: confusion from counting copies. Am J Gastroenterol 105:360–362
15. Armour JAL, Palla R, Zeeuwen PLJM et al (2007) Accurate, high-throughput typing of copy number variation using paralogue ratios from dispersed repeats. Nucleic Acids Res 35:e19–e19
16. Field SF, Howson JM, Maier LM et al (2009) Experimental aspects of copy number variant assays at CCL3L1. Nat Med 15:1115–1117
17. Aldhous MC, Bakar SA, Prescott NJ et al (2010) Measurement methods and accuracy in copy number variation: failure to replicate associations of beta-defensin copy number with Crohn's disease. Hum Mol Genet 19:4930–4938

18. Fode P, Jespersgaard C, Hardwick RJ et al (2011) Determination of beta-defensin genomic copy number in different populations: a comparison of three methods. PLoS One 6, e16768
19. Haridan US, Mokhtar U, Machado LR et al (2015) A comparison of assays for accurate copy number measurement of the low-affinity Fc gamma receptor genes FCGR3A and FCGR3B. PLoS One 10, e0116791
20. Deutsch S, Choudhury U, Merla G et al (2004) Detection of aneuploidies by paralogous sequence quantification. J Med Genet 41:908–915
21. Gilliland G, Perrin S, Blanchard K et al (1990) Analysis of cytokine mRNA and DNA: detection and quantitation by competitive polymerase chain reaction. Proc Natl Acad Sci U S A 87:2725–2729
22. Diviacco S, Norio P, Zentilin L et al (1992) A novel procedure for quantitative polymerase chain reaction by coamplification of competitive templates. Gene 122:313–320
23. Hollox EJ, Detering JC, Dehnugara T (2009) An integrated approach for measuring copy number variation at the FCGR3 (CD16) locus. Hum Mutat 30:477–484
24. Fernando MM, Boteva L, Morris DL et al (2010) Assessment of complement C4 gene copy number using the paralog ratio test. Hum Mutat 31:866–874
25. Hardwick RJ, Ménard A, Sironi M et al (2014) Haptoglobin (HP) and Haptoglobin-related protein (HPR) copy number variation, natural selection, and trypanosomiasis. Hum Genet 133:69–83
26. Khan FF, Carpenter D, Mitchell L et al (2013) Accurate measurement of gene copy number for human alpha-defensin DEFA1A3. BMC Genomics 14:719
27. Polley S, Louzada S, Forni D et al (2015) Evolution of the rapidly-mutating human salivary agglutinin gene (DMBT1) and population subsistence strategy. P Proc Natl Acad Sci U S A 112:5105–5110
28. Walker S, Janyakhantikul S, Armour JA (2009) Multiplex Paralogue Ratio Tests for accurate measurement of multiallelic CNVs. Genomics 93:98–103
29. Carpenter D, Walker S, Prescott N et al (2011) Accuracy and differential bias in copy number measurement of CCL3L1 in association studies with three auto-immune disorders. BMC Genomics 12:418
30. Saldanha G, Potter L, Dyall L et al (2011) Detection of copy number changes in DNA from formalin fixed paraffin embedded tissues using paralogue ratio tests. Anal Chem 83:3484–3492
31. Veal CD, Reekie KE, Lorentzen JC et al (2013) A 129 kb Deletion on Chromosome 12 Confers Substantial Protection Against Rheumatoid Arthritis, Implicating the Gene SLC2A3. Hum Mutat 35:248–256
32. Koontz D, Baecher K, Kobrynski L et al (2014) A pyrosequencing-based assay for the rapid detection of the 22q11.2 deletion in DNA from buccal and dried blood spot samples. J Mol Diagn 16:533–540
33. Carpenter D, Dhar S, Mitchell LM et al (2015) Obesity, starch digestion and amylase: association between copy number variants at human salivary (AMY1) and pancreatic (AMY2) amylase genes. Hum Mol Genet 24:3472–3480
34. Hallast P, Balaresque P, Bowden GR et al (2013) Recombination dynamics of a human Y-chromosomal palindrome: rapid GC-biased gene conversion, multi-kilobase conversion tracts, and rare inversions. PLoS Genet 9:e1003666
35. Royo JL, Pascual-Pons M, Lupianez A et al (2015) Genotyping of common SIRPB1 copy number variant using Paralogue Ratio Test coupled to MALDI-MS quantification. Mol Cell Probes 29:517–521
36. Hollox EJ, Huffmeier U, Zeeuwen PL et al (2008) Psoriasis is associated with increased beta-defensin genomic copy number. Nat Genet 40:23
37. Stuart PE, Hüffmeier U, Nair RP et al (2012) Association of β-defensin copy number and psoriasis in three cohorts of European origin. J Invest Dermatol 132:2407–2413
38. Perne A, Zhang X, Lehmann L et al (2009) Comparison of multiplex ligation-dependent probe amplification and real-time PCR accuracy for gene copy number quantification using the beta-defensin locus. Biotechniques 47:1023–1028
39. Hardwick RJ, Machado LR, Zuccherato LW et al (2011) A worldwide analysis of beta-defensin copy number variation suggests recent selection of a high-expressing DEFB103 gene copy in East Asia. Hum Mutat 32:743–750
40. Veal CD, Xu H, Reekie K et al (2013) Automated design of paralogue ratio test assays for the accurate and rapid typing of copy number variation. Bioinformatics 29:1997–2003
41. Kuhn RM, Haussler D, Kent WJ (2013) The UCSC genome browser and associated tools. Brief Bioinform 14:144–161
42. Bailey JA, Yavor AM, Massa HF et al (2001) Segmental duplications: organization and impact within the current human genome project assembly. Genome Res 11:1005–1017

43. Tempel S (2012) Using and understanding RepeatMasker. Methods Mol Biol 859:29–51
44. Redon R, Ishikawa S, Fitch KR et al (2006) Global variation in copy number in the human genome. Nature 444:444–454
45. Conrad DF, Pinto D, Redon R et al (2009) Origins and functional impact of copy number variation in the human genome. Nature 464:704–712
46. Sudmant PH, Kitzman JO, Antonacci F et al (2010) Diversity of human copy number variation and multicopy genes. Science 330:641
47. Sudmant PH, Rausch T, Gardner EJ et al (2015) An integrated map of structural variation in 2,504 human genomes. Nature 526:75–81
48. Wain LV, Odenthal-Hesse L, Abujaber R et al (2014) Copy number variation of the beta-defensin genes in europeans: no supporting evidence for association with lung function, chronic obstructive pulmonary disease or asthma. PLoS One 9, e84192
49. Barnes C, Plagnol V, Fitzgerald T et al (2008) A robust statistical method for case–control association testing with copy number variation. Nat Genet 40:1245–1252
50. Gonzalez JR, Subirana I, Escaramis G et al (2009) Accounting for uncertainty when assessing association between copy number and disease: a latent class model. BMC Bioinformatics 10:172
51. Subirana I, Diaz-Uriarte R, Lucas G et al (2011) CNVassoc: Association analysis of CNV data using R. BMC Med Genome 4:47
52. Cardin N, Holmes C, Wellcome Trust Case Control C et al (2011) Bayesian hierarchical mixture modeling to assign copy number from a targeted CNV array. Genet Epidemiol 35:536–548
53. Veal CD, Freeman PJ, Jacobs K et al (2012) A mechanistic basis for amplification differences between samples and between genome regions. BMC Genomics 13:455
54. Ottolini B, Hornsby MJ, Abujaber R et al (2014) Evidence of convergent evolution in humans and macaques supports an adaptive role for copy number variation of the beta-defensin-2 gene. Genome Biol Evol 6:3025–3038
55. Zhang X, Muller S, Moller M et al (2014) 8p23 beta-defensin copy number determination by single-locus pseudogene-based paralog ratio tests risk bias due to low-frequency sequence variations. BMC Genomics 15:64
56. Abu Bakar S, Hollox EJ, Armour JAL (2009) Allelic recombination between distinct genomic locations generates copy number diversity in human β-defensins. Proc Natl Acad Sci U S A 106:853–858
57. Gaunt TR, Rodriguez S, Guthrie PAI et al (2010) An expectation–maximization program for determining allelic spectrum from CNV data (CoNVEM): insights into population allelic architecture and its mutational history. Hum Mutat 31:414–420
58. Aklillu E, Odenthal-Hesse L, Bowdrey J et al (2013) CCL3L1 copy number, HIV load, and immune reconstitution in sub-Saharan Africans. BMC Infect Dis 13:536

Chapter 9

Genotyping Multiallelic Copy Number Variation with Multiplex Ligation-Dependent Probe Amplification (MLPA)

Suzan de Boer and Stefan J. White

Abstract

Multiallelic copy number variants are genomic loci that can be present in a range of different copy numbers between individuals. High or low copy numbers of specific genes have been associated with different diseases. Precise genotyping of these loci can be complicated, and relies on accurate assays. Multiplex ligation-dependent probe amplification (MLPA) is a PCR-based approach that allows copy number determination of up to 50 genomic loci in a single reaction. In this chapter, we outline the basic protocol, with a particular emphasis on the appropriate approach to accurately genotype multiallelic copy numbers.

Key words Copy number variation, Deletion, Duplication, Capillary electrophoresis, MLPA, PCR

1 Introduction

There are many types of genetic variation, in the human genome. One class is copy number variation (CNV), defined as a gain or a loss compared to the reference genome. A number of loci show a wide range of copy numbers between individuals, which collectively are known as multiallelic copy number variants, or mCNV. A number of different methodologies have been applied to the analysis of mCNV (reviewed in [1]). One approach is multiplex ligation-dependent probe amplification (MLPA), a PCR-based technique first described in 2002 [2].

MLPA is based around the ligation of two half probes which recognize a specific sequence of interest (Fig. 1). Ligation will only occur when both half probes are hybridized to their target sequences, and only ligated probes are amplified simultaneously during the PCR reaction. Because the probes contain identical ends, the ligated products can be amplified together with a single primer pair. One of the two primers in the PCR is fluorescently labeled, meaning that the amplified products can be visualized during fragment separation by capillary electrophoresis. Each probe is

Stefan J. White and Stuart Cantsilieris (eds.), *Genotyping: Methods and Protocols*, Methods in Molecular Biology, vol. 1492,
DOI 10.1007/978-1-4939-6442-0_9, © Springer Science+Business Media New York 2017

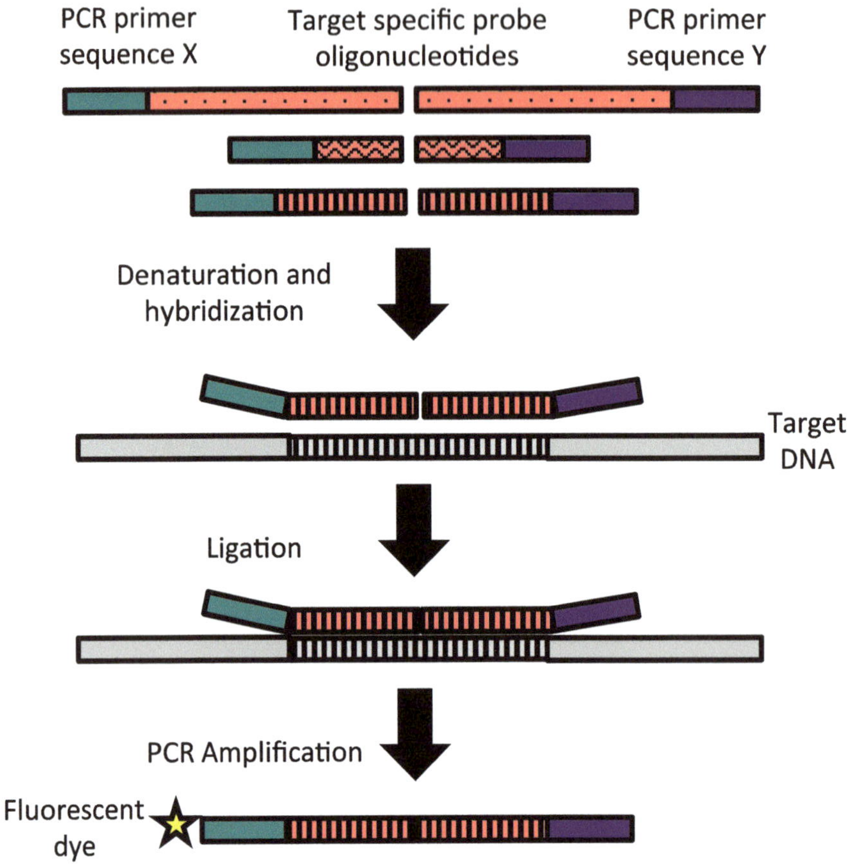

Fig. 1 The basis of MLPA. Genomic DNA is denaturized, with the half probes hybridizing to the single-stranded DNA. Only half probes that hybridize adjacently can be ligated together, and only ligated products can be amplified with PCR

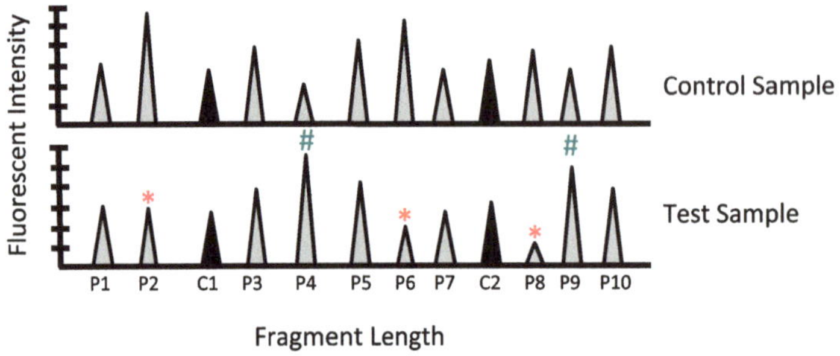

Fig. 2 Fragment separation by capillary electrophoresis distinguishes probes by their unique length, and will generate a consistent peak pattern. The peak heights of control probes (C1, C2) should be consistent between samples. Comparing relative differences in test probes (P1–P10) between the control sample and the test sample shows a decrease in copy number for probes P2, P6, and P8 (marked with *asterisk*), and a gain in copy number for probes P4 and P9 (marked with *hash*)

designed to have a unique length, and relative differences in peak heights correspond to changes in copy number (Fig. 2).

The principle advantages of MLPA are that it allows for a rapid (results being available within 24 h) and high-throughput quantification (96 samples can be handled simultaneously) of up to 50 sequences per DNA sample in a single reaction, by using a single PCR primer pair. It has been adapted to a range of different applications, including gene expression [3] and methylation analysis [4].

2 Materials

2.1 MLPA Reagents

All reagents for this MLPA protocol can be purchased from MRC-Holland, The Netherlands (www.mlpa.com). The different components can be recognized by a distinguishing cap color.

SALSA MLPA buffer (yellow cap)

SALSA Ligase-65 (green cap)

Ligase Buffer A (transparent cap)

Ligase Buffer B (white cap)

SALSA Polymerase (orange cap)

SALSA PCR Primer Mix (brown cap)

The PCR Primer Mix contains the following primers

Forward 5′-GGGTTCCCTAAGGGTTGGA-3′

Reverse 5′-GTGCCAGCAAGATCCAATCTAGA-3′

The forward primer is fluorescent labeled at the 5′ end, usually with FAM.

Probe mix (black cap)

MRC has a variety of ready-to-order probe mixes (black cap), or homemade probes can be developed using synthetic oligonucleotides [5]. When designing probes, the CG content of the hybridizing sequence of each half probe should be 35–60%, and the T_m should be greater than 66 °C. Finally, to maximize signal strength, the first nucleotide of the unique sequence of the left half probe should be a C or a G. The right oligonucleotide should be phosphorylated at the 5′ end, to allow ligation to take place.

MLPA probes are typically designed against unique sequences in the reference genome. Polymorphic loci are usually represented more than once, so extra care must be taken when choosing probe sequences. If a class of homologous genes is to be assayed, then it is important to choose sequences that are identical across all genes. Conversely, if a specific gene is to be studied then the oligonucleotides should be chosen such that any sequence mismatches are at or near the ligation site. Although a single mismatch may be sufficient to generate a specific product, it is preferable for multiple nucleotides to be different.

As probes are typically separated by capillary electrophoresis, it is essential that each probe has a different length. We have successfully used probes generating products within the size range of 80–150 bp.

2.2 Additional Materials and Reagents

Thermocycler with heated lid

Filter tips

PCR strip tubes with individual lids

Hi-Di Formamide (Applied Bioscience)

Size standard (Applied Bioscience)

3 Methods

The MLPA protocol below is an updated version of that described in the original publication [2]. It is also available at the website of MRC-Holland (www.mlpa.com).

1. Add 20–500 ng genomic DNA in a final volume of 5 μl to a PCR tube (*see* **Note 1**).
2. The DNA is denatured for 5 min at 98 °C, and should be allowed to cool to room temperature for at least 5 min (*see* **Note 2**).
3. To the genomic DNA add 1.5 μl MLPA probe mix and 1.5 μl SALSA MLPA buffer, and carefully mix. Incubate for 1 min at 95 °C, then 16 h at 60 °C (*see* **Note 3**).
4. Prepare the ligase mix at room temperature. Mix 3 μl Ligase-65 buffer A and 3 μl Ligase-65 buffer B in 25 μl H_2O. Add 1 μl Ligase-65 and mix again.
5. Reduce the temperature of the thermal cycler to 54 °C. While keeping the PCR tubes in the thermal cycler, add 32 μl of the ligase mix to each tube and mix (*see* **Note 4**). Incubate the reaction for 10–15 min at 54 °C, followed by 5 min at 98 °C to inactivate the ligase (*see* **Note 5**).
6. To make the polymerase master mix, prepare the following for each reaction (*see* **Note 6**):

H_2O	7.5 μl
SALSA PCR primer mix	2 μl
SALSA Polymerase	0.5 μl

7. Store the master mix on ice until use. At room temperature, add 10 μl polymerase master mix to each tube containing the MLPA ligation reaction and mix by gently pipetting.
8. Place the tubes in the thermocycler and run the PCR reaction with the following settings:

   1 cycle: 1 min 95 °C

   35 cycles: 30 s 95 °C; 30 s 60 °C; 30 s 72 °C

   1 cycle: 20 min 72 °C
9. Prepare samples for fragment analysis on a capillary sequencer (*see* **Note 7**). Add 5 μl size standard to 1 ml Hi Di Formamide and mix. Into each well of a 96-well plate, add 9 μl of the Formamide/size standard mix, and then add 1 μl of PCR product to each well (*see* **Note 8**).
10. Data analysis. Fragment separation is usually performed on a capillary sequencer, which measures absolute fluorescence. Peaks generated by capillary sequencing require normalization, which consist of two steps. First there is intrasample normalization, where the height of each probe peak is compared to the peak heights of reference probes within a sample to produce a ratio. An intersample normalization is then performed, by dividing each probe ratio by the median value of the matching probe ratios across all samples.

    For typical diploid loci this normalized ratio will be 1.0, with deleted and duplicated loci within individual samples having normalized ratios of ~0.5 and 1.5 respectively. When analyzing mCNV loci this will not be the case. There have been different approaches described for assigning specific copy numbers to samples when a range of copy numbers is expected. For high-quality data, it may be possible to identify distinct groups by eye. The copy number of each group can then be estimated by determining the proportional difference between the groups (Fig. 3). Copy number grouping can be improved by having multiple probes per locus, and using the average value [6, 7]. For less clear data it is possible to bin samples into arbitrary groups based on predefined borders, however this has the chance of introducing bias.

4 Notes

1. High-quality DNA, isolated in a consistent manner, is essential for a successful MLPA analysis. A degree of degradation can be tolerated, as the DNA sequence used as template for oligonucleotide hybridization is usually <200 bp. Sites of DNA breakage are unlikely to be completely random, however, meaning that a commonly degraded locus may appear as a (somatic)

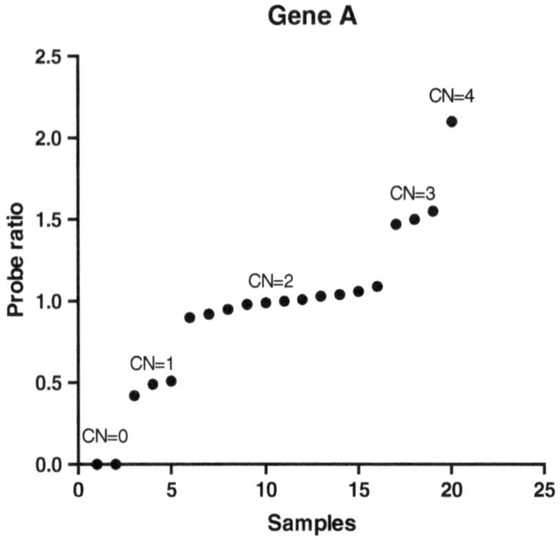

Fig. 3 Tight clustering of normalized ratios for each locus simplifies data interpretation. In this case the average difference between groups is 0.5, supporting the conclusion that the copy numbers (CN) range from 0 to 4

deletion. Impurities such as phenol can influence the MLPA reaction. The method in which the DNA is isolated and purified can have a subtle impact on relative peak heights, meaning that otherwise high-quality DNA samples, prepared using different protocols, may give spurious copy number differences. For this reason, it is best for all DNA samples within a study to be isolated using the same procedure. Similarly, if two or more study populations are being analyzed, the samples should be randomized during the MLPA procedure, rather than processing the populations in separate batches on different days.

2. Incomplete denaturation of the DNA can lead to reduced probe access, which can be resolved by increasing the denaturation time.
3. Successful use of a shorter hybridization step (2–3 h) has been reported [8].
4. The ligation step needs to be performed at 54 °C, therefore keep tubes in the thermocycler at 54 °C when adding the ligation mix.
5. Products are stable when stored at −20 °C.
6. The fluorescent labels are light-sensitive. Minimize exposure of the primer mix to light during the pipetting steps, and PCR products should be stored in the dark.
7. This step is for the ABI3700 capillary sequencer (Applied Biosystems).
8. It may be necessary to alter the amount of PCR product added to achieve optimal peak heights.

Acknowledgements

SB is supported by the Australian Postgraduate Award (APA) and the International Postgraduate Research Scholarship scholarships from Monash University.

References

1. Cantsilieris S, Baird PN, White SJ (2013) Molecular methods for genotyping complex copy number polymorphisms. Genomics 101:86–93
2. Schouten JP, McElgunn CJ, Waaijer R et al (2002) Relative quantification of 40 nucleic acid sequences by multiplex ligation-dependent probe amplification. Nucleic Acids Res 30:e57
3. Eldering E, Spek CA, Aberson HL et al (2003) Expression profiling via novel multiplex assay allows rapid assessment of gene regulation in defined signalling pathways. Nucleic Acids Res 31:e153
4. Nygren AO, Ameziane N, Duarte HM et al (2005) Methylation-specific MLPA (MS-MLPA): simultaneous detection of CpG methylation and copy number changes of up to 40 sequences. Nucleic Acids Res 33:e128
5. den Dunnen JT, White SJ (2006) MLPA and MAPH: sensitive detection of deletions and duplications. Curr Protoc Hum Genet 7:1–14
6. Groth M, Szafranski K, Taudien S et al (2008) High-resolution mapping of the 8p23.1 beta-defensin cluster reveals strictly concordant copy number variation of all genes. Hum Mutat 29:1247–1254
7. White SJ, Vissers LE, Geurts van Kessel A et al (2007) Variation of CNV distribution in five different ethnic populations. Cytogenet Genome Res 118:19–30
8. Aten E, White SJ, Kalf ME et al (2008) Methods to detect CNVs in the human genome. Cytogenet Genome Res 123:313–321

Chapter 10

Analysis of Multiallelic CNVs by Emulsion Haplotype Fusion PCR

Jess Tyson and John A.L. Armour

Abstract

Emulsion-fusion PCR recovers long-range sequence information by combining products in *cis* from individual genomic DNA molecules. Emulsion droplets act as very numerous small reaction chambers in which different PCR products from a single genomic DNA molecule are condensed into short joint products, to unite sequences in *cis* from widely separated genomic sites. These products can therefore provide information about the arrangement of sequences and variants at a larger scale than established long-read sequencing methods. The method has been useful in defining the phase of variants in haplotypes, the typing of inversions, and determining the configuration of sequence variants in multiallelic CNVs. In this description we outline the rationale for the application of emulsion-fusion PCR methods to the analysis of multiallelic CNVs, and give practical details for our own implementation of the method in that context.

Key words Emulsion, CNV, Haplotype, Phase, Structural variation, DNA sequencing

1 Introduction

Despite the spectacular successes of next-generation sequencing, some simple questions about genomic structure remain beyond the reach of standard approaches. In particular, even relatively deep accumulation of short reads is unable to resolve structural rearrangements *de novo*, or establish the long-range phase of linked variants. Even when paired ends of longer fragments are used to increase the range of coupling, if informative sites are separated at a distance greater than the fragment size, short reads can still fail to resolve the desired information. This can limit the determination of linkage phase when heterozygous sites are separated by a tract of homozygous DNA, or the establishment of structural variants when the boundaries of the rearrangement are bordered by segmental duplications (Fig. 1).

Where it is necessary to the investigation, several alternative approaches have been adopted to recover information about structure or phase on a larger scale than can be covered by short-read

Stefan J. White and Stuart Cantsilieris (eds.), *Genotyping: Methods and Protocols*, Methods in Molecular Biology, vol. 1492,
DOI 10.1007/978-1-4939-6442-0_10, © Springer Science+Business Media New York 2017

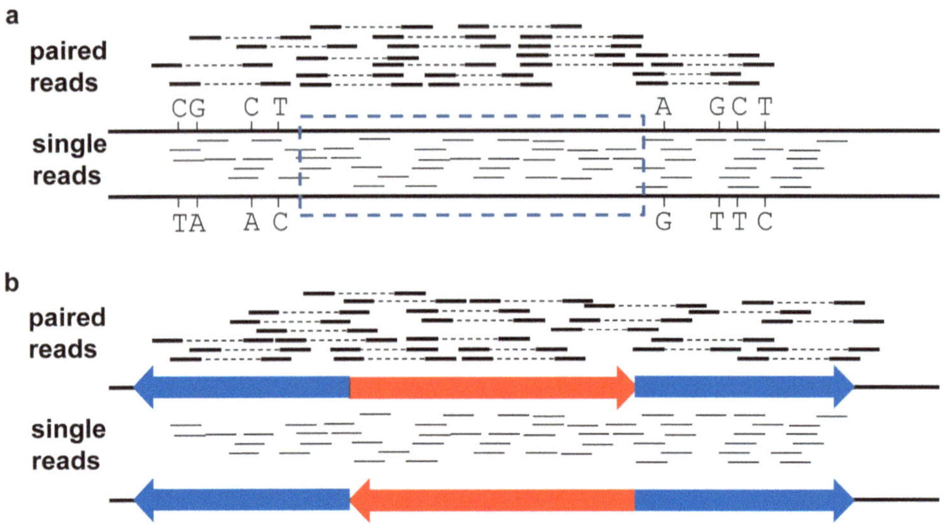

Fig. 1 Failure of short reads to document (**a**) phase of linked variants and (**b**) structural variation. (**a**) Short reads can encompass two linked variants if they are very close, or paired-end reads can establish phase over longer ranges, but a large area without informative variants (*dashed central box*) means that the phase between the two (*left* and *right*) groups of variants positions cannot be established. (**b**) In this example, the two haplotypes differ by an inversion (*red*) flanked by (*inverted*) segmental duplications (*blue*). Neither individual reads nor paired ends span the segmental duplications; as a result, no new structures are created on the scale of single or paired reads, and the two inversion alleles cannot be distinguished

sequencing. Optical mapping approaches, including fiber-FISH and nanochannel [1] (BioNano Genomics) methods, can be very useful for long-range structural determination, but do not systematically recover full sequence information. Large-insert (fosmid) cloning has been very powerful in providing gold-standard assemblies of genomes that faithfully preserve phase [2, 3], and long-read DNA sequencing technologies in both the SMRT [4] and nanopore [5] implementations are increasing in power and effectiveness. However, both cloning and long-read approaches provide whole-genome information, while what is often needed by investigators is targeted information in specific regions. In principle, employing long-read methods or cloning on isolated locus-specific DNA could provide an efficient solution, but such a method would depend on the development of a successful technology to enrich long double-stranded DNA for regions of specific interest. At present, locus-specific enrichment methods either result in short fragments or depend on long PCR, which can be subject to rearrangement by in vitro recombination.

In this chapter we describe practical methods for emulsion-fusion PCR, as a method to recover the information in single genomic DNA molecules from specific genomic regions. The core of the method is to combine fusion PCR, to fuse two coamplified molecules into a single-joint product, with separation of reactions into numerous emulsion droplets, so that each droplet contains only the products from a single original target molecule. As a result, each joint product molecule contains the corresponding components fused in *cis*, preserving the phase of the original longer region. This approach has been applied to the determination of spatial relationships between DNA molecules, for example in inversions, and to establish the phase of linked variants [6–9]. A detailed explanatory protocol for those applications to single-copy sequences has been published elsewhere [10], and this chapter concentrates on methods we have developed appropriate to the resolution of spatial information in multiallelic CNVs. Our modifications to the established methods include the resolution of multiple-fusion products by allele-specific PCR, and the inclusion of longer fusion partners, so that the required information can be efficiently recovered as direct sequencing data, rather than by more indirect genotyping methods. In the research that prompted these developments, our primary aim was to couple sequence variation in copy-variable repeats with spatial information: Where CNV repeats differ by sequence variation, where is each sequence variant (Fig. 2)? This question subdivides into the primary question of which repeat variants are present on each of the two haplotypes, and the secondary question of how the repeats in each haplotype are arranged along the array [11, 12].

In practice, we found that at the human alpha-defensin CNV, which has a 19 kb repeat unit, emulsion-fusion PCR could place sequence variants reliably in the first two repeat units in from either end of the array, but longer range fusion products were not detectable [12]. Although we will describe the methods for application to a multiallelic CNV such as human alpha-defensin, these same methods are of course transferrable to simpler questions of phase or structure in single-copy regions. Setting up a successful emulsion-fusion system involves three specialized steps: specific amplicon design, primary emulsion-fusion PCR, and secondary selective PCRs to prepare templates for direct sequencing.

1.1 Primer Design and Preliminary Testing

The main protocol and comments below will cover the use of primers in the emulsion-fusion reaction, but careful attention to primer design and testing is crucial to the success of the experiment, and is outlined in this section. In the protocol that follows, we name the primers so that primers F1 and R1 amplify amplicon 1, and primers F2 and R2 amplify amplicon 2. Product size can range from a few hundred base pairs to 2 kb. Before attempting to undertake emulsion-fusion with these amplicons, first design them so that they follow the usual rules for genomic amplicons (avoidance of dispersed repeats, primer SNPs, etc.), with the added condition that the F1/R1 and F2/R2 pairs should both work under the same conditions. A certain amount of forethought is needed to make sure that the

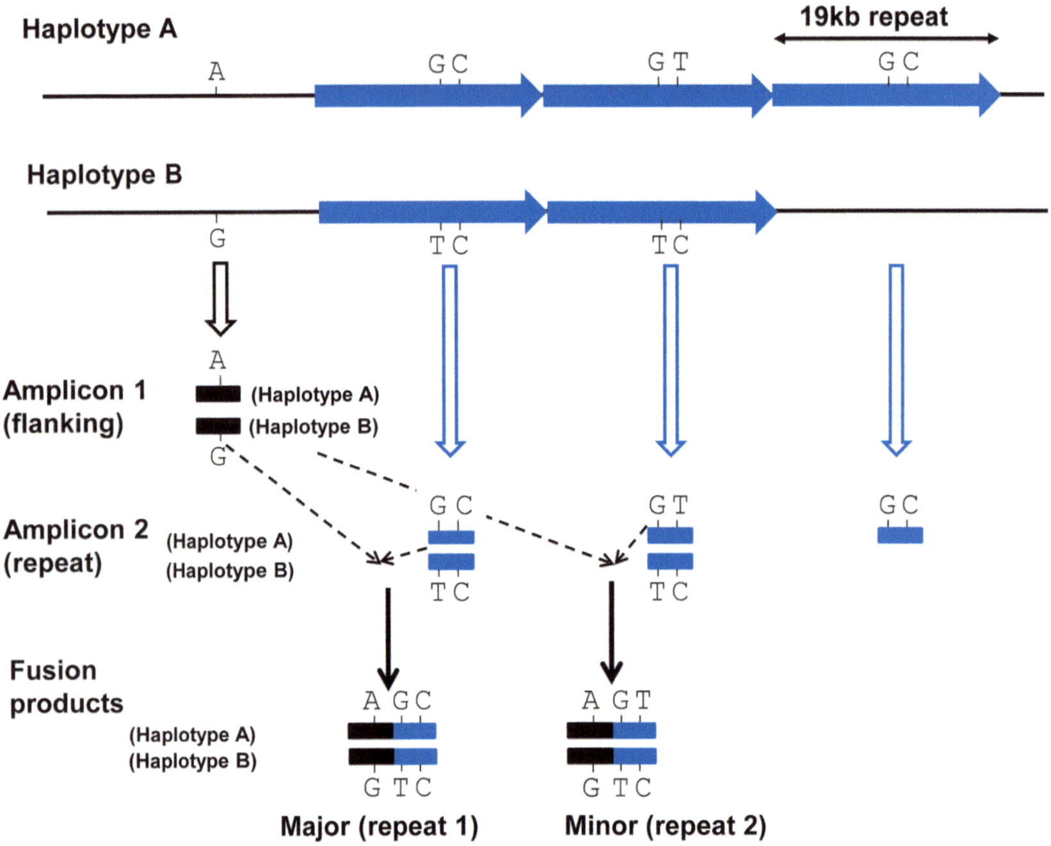

Fig. 2 Summary of emulsion-fusion analysis at a multiallelic CNV. The analysis shown investigates a five-copy individual at the human alpha-defensin CNV, which has a 19 kb repeat unit. The two haplotypes A (three repeat units) and B (two repeat units) are shown, with two internal positions per repeat at which these repeats differ in sequence; an informative SNP in the flanking DNA is also shown. The two amplicons encompass the informative SNP in the flanking DNA (amplicon 1) or the variable positions in the repeat unit (amplicon 2). After emulsion-fusion PCR, fusion in *cis* between amplicons 1 and 2 generates major products from the nearest repeat, and minor products from the second repeat. The mixture of products can be resolved, and the spatial arrangement of repeat variants reconstructed, after allele-specific PCR and sequencing. Allele-specific PCR specific for flanking A will selectively amplify products from haplotype A, and sequencing will differentiate a major (GC) and a minor (GT) sequence profile; allele-specific PCR specific for flanking G will selectively amplify products from haplotype B, and sequencing will display the TC sequence combination from both the first and second repeat units

fusion products will contain key informative positions in accessible locations (for example, if Sanger sequencing is to be used for the final readout, remember to allow 50–60 bases from the primer before useful sequence begins). Similarly, if the investigation is going to make use of allele-specific primers to separate the haplotypes on the basis of heterozygous positions, make sure that effective allele-specific primers can be designed for these informative positions. If using allele-specific primers, the ideal strategy is to make use of the first informative position in the flanking amplicon, so that any further flanking heterozygous sites can be used as validation of allele specificity. After emulsion-fusion, our experience is that clean products can only be reliably obtained if the initial fusion products are reamplified using nested primers. It is therefore wise to design the initial amplicons with allele-specific and nested primers in mind. Once these have been designed, test each amplicon pair individually, to verify that they each gives good yields of clean products under the same cycling conditions; if necessary, modify the primers until they do. Although F2 will not be used in the fusion reaction, the 3′ end of products made with the fusion primer F2'R1 will be expected to have the same properties as F2.

The key to the emulsion-fusion method we describe here is the (single) central fusion primer F2'R1 [10]. As its name implies, this is a composite primer formed of (at its 5′ end) the complement of primer F2, and (at its 3′ end) primer R1. In the simplest fusion reaction (Fig. 3), the three primers F1, F2'R1, and R2 are used together. Use of F1 in conjunction with F2'R1 yields F1-R1 products with (on the top strand) a 3′ extension corresponding to the F2 sequence. These F2-extended F1-R1 top strands then act as megaprimers to initiate PCR with R2, thereby achieving the F1-[F2'R1]-R2 fusion. As an example of successful primers, here

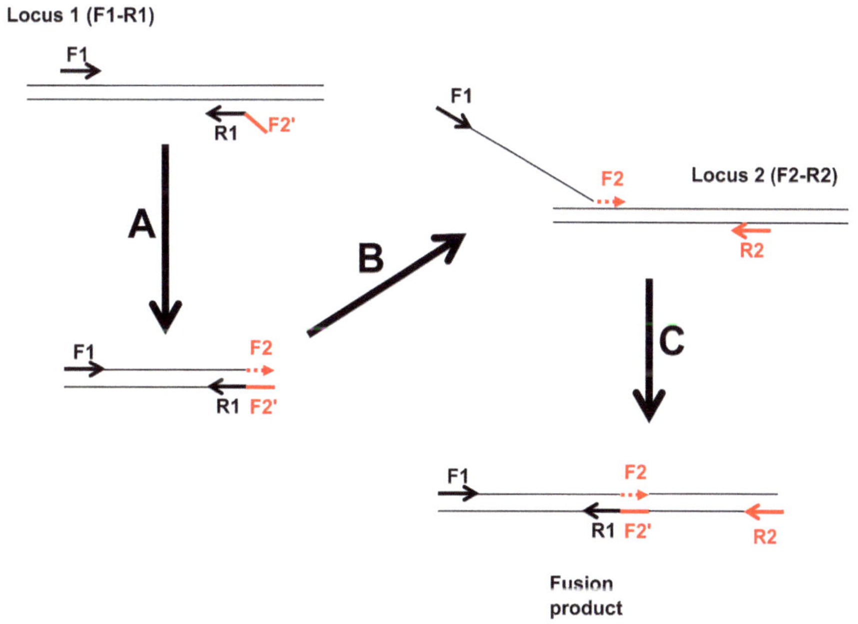

Fig. 3 The underlying process of PCR fusion between two amplicons. At locus 1, amplification proceeds between F1 and R1, and extension of the top strand to copy the F2' tail of the composite F2'R1 primer yields products from locus 1 with a top strand ending in the F2 primer (A). This F2 end of the modified amplicon 1 can now prime at the F2 site of locus 2 (B). The resulting fusion product can subsequently be amplified with primers F1 and R2 (C)

are the primers used to amplify alpha-defensin units by Tyson and Armour (2012); as emphasized above, only F1, F2'R1, and R2 are used in the fusion reactions, and primers R1 and F2 are only used in the preliminary testing stage [12]:

```
F1       TTCATTGGCCACCCTGGACT
[R1      TTCCCCCTGGAAACCTCAATCC]
[F2      TGCACGCTGGTATTCTGCAA]
R2       CGGGAGAGAGGTTCCAGAGTTG
F2'R1    TTGCAGAATACCAGCGTGCATTCCCCCTGGAAACCTCAATCC
          <-Complement of F2-><-    primer R1     ->
```

2 Materials

2.1 Preparation of Oil Phase

1. Silicone polyether/cyclopentasiloxane (Dow Corning).
2. Cyclopentasiloxane/trimethylsiloxysilicate (Dow Corning).
3. Silicone oil AR 20 (Sigma-Aldrich).
4. 3 mm Tungsten carbide beads (Qiagen).

2.2 Aqueous PCR

1. Phusion High-Fidelity DNA Polymerase (*see* **Note 1**).
2. LongAmp *Taq* DNA polymerase.
3. dNTPs.
4. Oligonucleotides, purified using the desalted option.

2.3 Disruption of Emulsion

1. Hexane.

2.4 Equipment

1. PCR thermocycler, capable of holding 0.2 or 0.5 ml PCR tubes.
2. Vortex Genie 2 for generation of emulsions (*see* **Note 2**).
3. Centrifuge, capable of speeds of ≥13,000 × *g* (*see* **Note 3**).

2.5 Primer Design

1. Primary PCR: Design two pairs of primers to independently amplify regions of interest to be fused, matching Tm to within ±1–2 °C. These primers are denoted F1 and R1 (locus 1 amplicon) and F2 and R2 for the locus 2 amplicon (*see* **Note 4**, and "Primer design and preliminary testing" above). The sequence corresponding to the reverse complement of the forward primer from locus 2 (F2) is appended to the reverse primer from locus 1 (R1) to make a tailed primer, F2′R1. Primers F1, F2′R1, and R2 are the three primers used in the emulsion PCR.

2. Secondary PCR: Design a single pair of nested primers (denoted F1N and R2N) to reamplify the fused product. One of these nested primers can be allele specific, so that the subset of products from a specific haplotype is amplified.

3 Methods

1. To make the oil phase; add 20 ml silicone polyether/cyclopentasiloxane, 15 ml cyclopentasiloxane/trimethylsiloxysilicate, and 15 ml silicone oil AR 20, in a 50 ml Falcon tube, inverting gently to mix. Store at room temperature. Mix gently by inversion before each use.
2. Prepare 25 μl aqueous phase for each DNA and two controls. Typical concentrations are shown in Table 1. One control will have no DNA (serves as a blank) and the second control will include DNA, but will not be subjected to the first cycling reaction (*see* **Note 5**).
3. Aliquot the aqueous phase to 0.5 ml PCR tubes, and add 50 μl oil phase (*see* **Note 6**).
4. Add a 3 mm tungsten carbide bead to the lid of the tube and close the tube keeping it in an inverted position (*see* **Note 7**).
5. Vortex the inverted tube at speed 5 for 1 min 30 s using a Vortex Genie 2.
6. Centrifuge all tubes for a few seconds at 3000 rpm to ensure that the contents are at the base of the tube before performing the cycling reaction.
7. A typical PCR is as follows (*see* **Note 8**):

Table 1

Volume and concentration of aqueous phase for emulsion PCR

	Concentration	Volume (μl)	Final concentration
Genomic DNA	50 ng/μl	1	2 ng/μl
Phusion GC buffer	5×	5	1×
dNTPs	10 mM	0.5	200 μM of each
Primer F1	10 μM	2.5	1 μM
Primer F2′R1	1 μM	0.625	25 nM
Primer R2	10 μM	2.5	1 μM
Phusion polymerase	2 U/μl	1	2 U
dH_2O		11.875	

98 °C for 30 s and 40 cycles of 98 °C for 10 s, annealing for 30 s and 72 °C for 30–60 s (depending on length of PCR product and polymerase used; for Phusion polymerase, use 30 s per 1 kb in the extension step). This is followed by 72 °C for 5 min and a 4 °C hold (*see* **Note 9**).

8. Once the PCR is complete, visually inspect the PCR tube to ensure that separation of the emulsion has not occurred during the PCR (*see* **Note 10**). Aim to disrupt the emulsions as soon as possible after the final cycle of the PCR.
9. In a fume cupboard add 200 μl of hexane to each emulsion, pipette to remove this mixture to a fresh 0.5 ml tube (discarding the original PCR tube and 3 mm tungsten carbide bead), and add 25 μl of 1× Phusion GC buffer (NEB) to increase the volume of the aqueous phase (*see* **Note 11**).
10. Vortex the closed tube in the upright position for at least 30 s, ensuring that a homogenous mixture is made. Include the non-cycle emulsion control at this stage.
11. Centrifuge for 3 min at 13,000 × *g*, using a microfuge suitable for 0.5 ml tubes. Distinct layers should be formed. In the fume cupboard, remove the top layer carefully (so as not to disturb the interface) and discard.
12. Repeat **steps 9–11** for an additional two times.
13. Leave tubes in the fume cupboard, uncovered, for 15 min to allow any remaining hexane to evaporate. Primary emulsion products can be used immediately in **step 14** or stored at −20 °C.
14. Secondary amplification: *See* **Note 12** and Table 2. Prepare 25 μl aqueous phase for each primary PCR product (including the non-cycle emulsion control extracted alongside the emul-

Table 2

Typical composition of a secondary PCR

	Concentration	Volume (μl)	Final concentration
Primary PCR product from **step 13** (*see* **Note 14**)		1	
LongAmp *Taq* reaction buffer	5×	5	1×
dNTPs	10 mM	0.75	300 μM of each
Primer F1N	10 μM	1	0.4 μM
Primer R2N	10 μM	1	0.4 μM
LongAmp *Taq* DNA polymerase	2.5 U/μl	1	2.5 U
dH_2O		15.25	

sion products in **step 9**) and two additional controls: no DNA, and 10 ng genomic DNA (*see* **Note 13**). Conditions for secondary PCR are dependent on the manufacturer's instructions for a particular polymerase. For LongAmp taq DNA polymerase, typical cycling conditions are as follows: 94 °C for 30 s and 30 cycles of 94 °C for 30 s, annealing for 30 s and 65 °C for 50 s/kb. This is followed by 65 °C for 10 min.

15. Visualize PCR products on an agarose gel alongside a molecular weight DNA ladder, to confirm expected size.
16. For sequencing of the fused product, purify PCR products using a bead- or gel-based purification protocol following the manufacturer's protocol.

4 Notes

1. Other polymerases may be used, but must generate a blunt end.
2. The speed of vortexers will vary, and therefore it may be necessary to optimize the vortex step to determine the setting required to produce aqueous compartments of 5–10 μM in diameter. Diameters of aqueous compartments can be estimated by light microscopy.
3. This speed is essential for the precise separation of the aqueous and oil phases post-PCR.
4. An additional primer, F2A, is optional, but, in our experience, necessary for the generation of amplicons >1 kb at locus 2. F2A is used in the emulsion stage, and ensures that early in the process a double-stranded product is exponentially amplified from locus 2, thus removing the reliance on linear synthesis from the R2 primer. F2A must be unique and set back from the F2 sequence, rather than using the F2 sequence itself, to avoid two complementary sequences (F2′R1 and F2) interfering with each other directly in the emulsion PCR. For amplicons at locus 2 >1 kb in length, use 0.25 μl F2A (10 μM) per reaction, adjusting the volume of dH_2O.
5. This will be the non-cycle emulsion control, used in the second round of PCR, to demonstrate that a fused product will only be obtained following two rounds of PCR.
6. The oil phase is viscous; leave the pipette tip in the oil for a few seconds to ensure that the intended volume has been added.
7. Generating the emulsion in the inverted orientation is essential for emulsion formation. Confining the bead in the conical base of the tube hinders the physical action of the bead in making the emulsion. If this happens, the emulsion will not be properly formed and may separate during the PCR.

8. For all emulsion experiments, it is important to establish PCR conditions that give robust yields of pure products in solution prior to being carried out in an emulsion.
9. If the thermocycler allows it, set to "safe" mode (to allow for the larger PCR volumes than usual) or set the sample volume to 75 μl.
10. If any separation of the emulsion has occurred during PCR this will be visualized as clear liquid at the base of the tube near the bead. Do not carry over any of the separated emulsion. Any separation means that fusion PCR in *trans* can occur in the aqueous phase and the product may not be derived from a single starting molecule.
11. This aids recovery and means no further dilution of this product is required for the second round of PCR.
12. Amplification of the fused product is only observed after two rounds of PCR. It is at this second stage that the use of at least one nested primer ensures a specific secondary product. In our experience, a predominance of unwanted products of incorrect size is observed if the same pair of primers is used in both the emulsion PCR stage and the secondary amplification.
13. A lack of amplification from these controls demonstrates that the product is dependent on both rounds of PCR and that the cycling conditions used do not permit the formation of the fused product in the absence of these steps.
14. The interface may still be visible even after three hexane extractions. For the secondary PCR, remove 1 μl of this product from the base of the tube to avoid the interface layer.

References

1. Hastie AR, Dong L, Smith A et al (2013) Rapid genome mapping in nanochannel arrays for highly complete and accurate *de novo* sequence assembly of the complex *Aegilops tauschii* genome. PLoS One 8:e55864
2. Duitama J, McEwen GK, Huebsch T et al (2012) Fosmid-based whole genome haplotyping of a HapMap trio child: evaluation of Single Individual Haplotyping techniques. Nucleic Acids Res 40:2041–2053
3. Suk E-K, McEwen GK, Duitama J et al (2011) A comprehensively molecular haplotype-resolved genome of a European individual. Genome Res 21:1672–1685
4. Huddleston J, Ranade S, Malig M et al (2014) Reconstructing complex regions of genomes using long-read sequencing technology. Genome Res 24:688–696
5. Cherf GM, Lieberman KR, Rashid H et al (2012) Automated forward and reverse ratcheting of DNA in a nanopore at 5-A precision. Nat Biotechnol 30:344–348
6. Wetmur JG, Chen J (2011) Linking emulsion PCR haplotype analysis. Methods Mol Biol 687:165–175
7. Wetmur JG, Kumar M, Zhang L et al (2005) Molecular haplotyping by linking emulsion PCR: analysis of paraoxonase 1 haplotypes and phenotypes. Nucleic Acids Res 33:2615–2619
8. Turner DJ, Shendure J, Porreca G et al (2006) Assaying chromosomal inversions by single-molecule haplotyping. Nat Methods 3:439–445
9. Turner DJ, Tyler-Smith C, Hurles ME (2008) Long-range, high-throughput haplotype determination via haplotype-fusion PCR and ligation haplotyping. Nucleic Acids Res 36:e82
10. Turner DJ, Hurles ME (2009) High-throughput haplotype determination over long distances by haplotype fusion PCR and ligation haplotyping. Nat Protoc 4:1771–1783

11. Black H, Khan F, Tyson J et al (2014) Inferring mechanisms of copy number change from haplotype structures at the human DEFA1A3 locus. BMC Genomics 15:614

12. Tyson J, Armour JAL (2012) Determination of haplotypes at structurally complex regions using emulsion haplotype fusion PCR. BMC Genomics 13:693

Chapter 11

Quantitative DNA Analysis Using Droplet Digital PCR

Rolf H.A.M Vossen and Stefan J. White

Abstract

Droplet digital PCR (ddPCR) is based on the isolated amplification of thousands of individual DNA molecules simultaneously, with each molecule compartmentalized in a droplet. The presence of amplified product in each droplet is indicated by a fluorescent signal, and the proportion of positive droplets allows the precise quantification of a given sequence. In this chapter we briefly outline the basis of ddPCR, and describe two different applications using the Bio-Rad QX200 system: genotyping copy number variation and quantification of Illumina sequencing libraries.

Key words Digital PCR, Copy number variation, DNA quantitation, NGS

1 Introduction

PCR typically involves the amplification of many DNA molecules in a single tube. In contrast, digital PCR divides individual DNA molecules into many parallel reactions. Initial applications used separate tubes or wells for each sub-reaction [1, 2], which is both cumbersome and costly for routine use. One solution to this problem is droplet digital PCR (ddPCR). By mixing a DNA sample with a water-oil mix under the right conditions, it is possible to generate thousands of droplets containing no, one, or more than one DNA molecules. Each droplet is an independent reaction that can be amplified simultaneously in a single tube, and only droplets that initially contained at least one DNA molecule will contain target product for detection.

1.1 ddPCR Reaction Chemistry

Target detection can be performed with dual-labeled hydrolysis probes (e.g., Taqman probes) or a dsDNA-binding dye like EvaGreen. Hydrolysis probes add more assay specificity and there is the possibility to measure more than one target in the same reaction by using probes that have different spectral wavelengths. Commonly used fluorescent dyes are FAM, VIC, and HEX. With probes, the unknown target and a control target are typically being measured in

Stefan J. White and Stuart Cantsilieris (eds.), *Genotyping: Methods and Protocols*, Methods in Molecular Biology, vol. 1492,
DOI 10.1007/978-1-4939-6442-0_11, © Springer Science+Business Media New York 2017

the same reaction. This also minimizes possible pipetting errors that otherwise could be introduced when the unknown target and control target are amplified in separate reactions.

The use of EvaGreen adds more flexibility because already available primers can be used. For small-scale experiments, the use of EvaGreen is a more cost-effective approach than dual-labeled probes. However, when a large number of DNA samples is going to be analyzed it may be more convenient to use probes. Determining, e.g., copy number variation (CNV) with EvaGreen assays usually means splitting up PCR reactions between the unknown target and a control target, thus doubling the number of reactions. Differentiating two size-different amplicons in a single reaction with EvaGreen has been described [3], with the longer amplicon giving a stronger signal than the shorter amplicon. Careful optimization of this assay is required to ensure sufficient discrimination between the two products.

1.2 Template Considerations

For optimal results, it is necessary to determine how much DNA is needed in a reaction. An input titration may be required to determine the optimal amount. Depending on the organism, genomic DNA templates may contain many fewer target copies than, e.g., cDNA or NGS libraries. Therefore, the total amount of input genomic DNA needed to detect the target copies will be much higher. For single-copy targets, one diploid genome is the equivalent of two target molecules. With NGS libraries, every molecule is a target molecule, as all molecules contain the same adapter sequences. As an example, 10 ng of human gDNA contains about 3000 copies of a unique sequence, whereas 10 ng of an Illumina library with an average size of 500 bp would contain roughly 1.86×10^{10} molecules. NGS libraries must therefore be diluted many times to achieve a sample concentration that is within the dynamic range of the system (100–5000 molecules/μl in the final emulsion reaction).

For genomic DNA samples being amplified in the QX200 it is best to measure the concentrations beforehand with devices like the Nanodrop (Thermo Scientific) or the Qubit (Invitrogen). Qubit is preferred since the Nanodrop can overestimate the concentration due to strong spectral absorption from contaminants or free nucleotides in solution.

For gene expression analysis, 1 ng of cDNA is a good starting point. Also here, depending on the gene expression levels, one may have to optimize the input amount.

1.3 Assay Design

When designing primers and probes for ddPCR, the same rules apply as for conventional rt-qPCR, with or without using a primer- or probe-design program. Standard desalted primers can be used and dual-labeled probes will usually be ordered as 20× concentrated assays. When designing primers for expression analysis, it is convenient to use Web-based programs like the Universal

ProbeLibrary Assay Design Center (accessible via lifescience.roche.com). For CNV assays, one must first map all restriction sites within the target sequence since the genomic DNA is being digested with a frequent cutter to ensure separation of all tandemly arranged gene copies. Digestion is performed in the PCR reaction. Just before thermal cycling, the appropriate restriction enzyme is added and reactions are incubated at room temperature.

1.4 Assay Optimization

For best results it is recommended that for each new target the annealing temperature is being optimized. This can be done by running a temperature gradient from 55 to 65 °C, as most targets will have an optimum temperature within that range when using

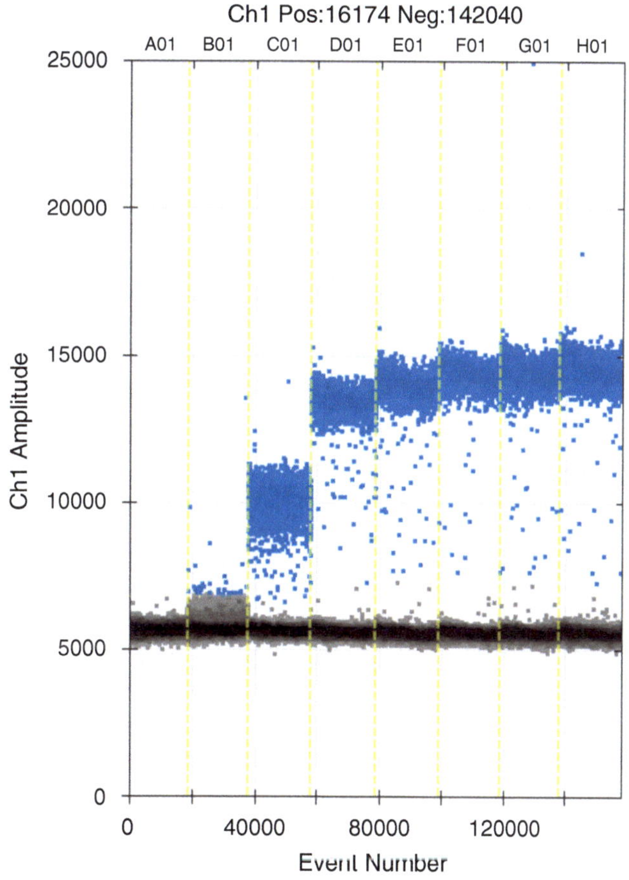

Fig. 1 Temperature gradient from 65 to 55 °C (from *left* to *right*, A01-H01) of a genomic single-copy target. The plot shows the FAM fluorescent intensity (Ch1 amplitude) at different temperatures. The temperature range where the largest separation is achieved between positive droplets (*upper clusters*) and negative droplets (*lower clusters*) is considered as optimal

Bio-Rad chemistry. The effect of temperature can be dramatic, as can be shown in Fig. 1.

The optimal annealing temperature for ddPCR Probe assays from Bio-Rad is 60 °C. The length of the amplicon can be kept short, e.g., 50–250 bp. With this amplicon size a two-step PCR with combined annealing/extension is used. For longer amplicons like 500–1500 bp, a three-step cycling program with a separate extension at 72 °C is preferred.

EvaGreen assays have optimal primer concentrations in the range of 100–200 nM final concentration. For probe assays the primers are at 900 nM and the probes at 250 nM final concentration.

1.5 Digital PCR Versus Conventional Real-Time qPCR

Real-time qPCR (rt-qPCR) has been routinely used since the early 1990s for different purposes, including measuring gene expression differences and quantitation of DNA. Although rt-qPCR is capable of detecting large differences in the number of target molecules between samples, it is less suitable for the detection of small differences. Quantitative differences in rt-qPCR are derived from the cycle of threshold (Ct), which is an exponential value. A twofold difference in the amount of target molecules means a Ct-value difference of 1, which is at the limit of where reproducible and accurate measurements can be taken. In practice, pipetting errors will often contribute to Ct value differences of up to one, and multiple PCR replicates (at least triplicates) are needed. Even with replicates, however, small differences will be measured with low accuracy. In rt-qPCR there may also be differences in PCR efficiency between primer pairs, an effect which has to be corrected for. Another important factor to consider is the heterogeneity of the target molecule population. For example, random next-generation sequencing (NGS) libraries contain a population of fragments that differ in size and sequence. In rt-qPCR, all these different fragments are being amplified simultaneously in the same reaction. An amplification bias is likely to occur, due to preferential amplification of smaller fragments. This, in turn, could lead to a false estimation of the original library quantity. Lastly, quantification with rt-qPCR is often inaccurate because it relies on comparing unknown samples with a standard curve. If the DNA concentration of the sample being used for the standard curve was not measured with an absolute method such as digital PCR, absolute quantification with rt-qPCR will not be possible. In contrast, digital PCR does not suffer from the abovementioned weaknesses, and since ddPCR is an end-point measurement, the effect of differences in PCR efficiency is of lesser importance. As such, digital PCR has been applied to a number of different applications where precision and/or sensitivity are required (reviewed in [4]).

1.6 Determining Copy Number Variation

The precision of ddPCR makes it a powerful approach for determining a wide range of copy numbers [5]. The amount of DNA needed for determining CNV depends on the expected highest target copy number. For targets with higher copy numbers, less DNA is required. If the expected copy number ranges from 1 to 10 then 10 to 60 ng input DNA is sufficient; for more then ten copy numbers <15 ng per reaction is needed. DNA samples are assayed with probes for the target gene of interest and a single copy reference gene, e.g., *RPP30* (NCBI Gene ID: 10556). The target copy number is calculated from the ratio of target to reference (*see* **Note 1**). Target and reference can be assayed separately in different reactions or together in the same reaction. Duplex detection is possible by using probes with a different fluorescent dye for each target (*see* **Note 2**).

To be sure that all target copies are separated, the DNA is digested in the PCR reaction prior to thermal cycling. To prevent digestion of the target gene, the sequence should be analyzed with a program like RestrictionMapper (www.restrictionmapper.org) to determine which enzymes can be used (*see* **Note 3**). During PCR reaction preparation, 2–5 U of restriction enzyme per well is added and incubated for 5–10 min at room temperature (RT).

1.7 Absolute Quantification of Illumina Libraries

Accurate quantification of Illumina NGS libraries is key to successful sequencing. While rt-qPCR is well suited for amplification of single fragments it is less optimal for the quantification of samples in which different fragments are amplified simultaneously. NGS libraries consist of a population of size- and sequence-different fragments that in qPCR may introduce amplification bias. At the point where the Ct value is measured, the fragment distribution of the library may have changed as the result of 10–20 amplification rounds. In contrast, ddPCR is an end-point measurement and it counts the individual molecules of the original library.

Libraries can be assayed either with probes or EvaGreen. A commercial library quantification kit for Illumina TruSeq libraries is available from Bio-Rad and is based on probes. With a probe assay one may be able to discriminate between adapter dimers and genuine library molecules. Adapter dimers have the smallest size and will amplify more efficiently than library molecules with larger insert sizes. As a result, a stronger fluorescent signal is generated from these molecules in a probe assay and a separate cluster of droplets containing only adapter dimers is often seen.

After Illumina library preparation, samples are often stored as 2 or 4 nM stock, as measured with a 2100 Bioanalyzer (Agilent) and/or Qubit Fluorometer (Thermo Fisher Scientific). For ddPCR, it is convenient to first make a 20 pM dilution from the stock (*see* **Note 3**).

Subsequently, several dilutions are made, from the 20 pM sample, e.g., 1000×, 2000×, and 3000×. Dilutions that give 100–5000 copies/μl can be used for quantification (Fig. 2). For samples that fall outside this range, additional dilutions have to be made.

Absolute library concentration can be calculated as follows:

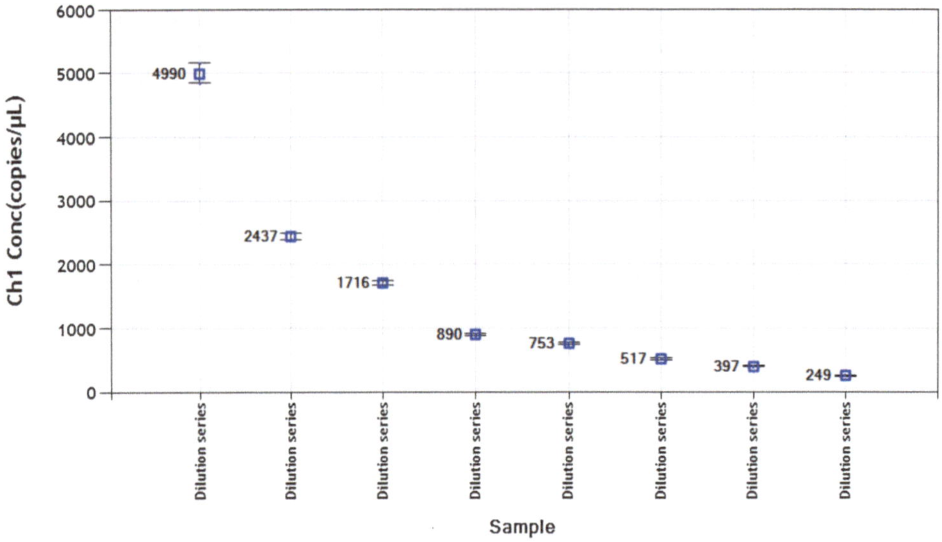

Fig. 2 Nonlinear serial dilution series of a 20 pM Illumina library (from *left* to *right*: 1000×–15,000×). Molecule counts range from 249 to 4990 copies/µl, and are all within the dynamic range of the system

Concentration (pM) = (copies/µl × dilution factor × dilution factor in reaction)/6 × 10^5

Example:

Suppose 1716 copies/µl were measured in a 3000× dilution, when 4 µl of DNA was initially added to a 22 µl mix. The concentration then would be

$(1716 \times 3000 \times 5.5)/6 \times 10^5 = 47.2$ pM

2 Materials

1. QX200 Droplet Digital PCR system with Automated Droplet Generator and Droplet Reader (Bio-Rad) including the QuantaSoft™ package.
2. T100 thermal cycler (Bio-Rad) or equivalent machine with adjustable ramp times.
3. ddPCR probe assays (Bio-Rad) or 20× TaqMan probe assays (Thermofisher Scientific).
4. Target PCR primers, standard desalted.
5. 2× ddPCR Supermix for probes (Bio-Rad).
6. 2× ddPCR EvaGreen Supermix (Bio-Rad).
7. Twin-Tec 96-well plates, semi-skirted (Eppendorf).
8. Droplet generation cartridges and gaskets (Bio-Rad).

9. Droplet generation oil for probes (Bio-Rad).
10. Droplet generation oil for EvaGreen (Bio-Rad).
11. Droplet Reader oil (Bio-Rad).
12. Pipette tips for AutoDG system (Bio-Rad).
13. Heat seals (Bio-Rad).
14. PX1 PCR plate sealer (Bio-Rad).
15. Diluting solution: 10 mM Tris–HCl pH 8, 0.05% Tween 20.
16. Restriction endonucleases such as HindIII or HaeIII (*see* **Note 4**).
17. Nanodrop Spectrophotometer (Thermo Scientific) or Qubit Fluorometer (Thermo Fisher Scientific).

3 Methods

The QX200 system measures absolute DNA quantities in 20 μl emulsion PCR reactions that contain 20,000 droplets. At our site the QX200 system is used for CNV, quantification of Illumina NGS libraries, and gene expression analysis (*see* **Note 5**).

3.1 Prepare PCR Reactions

1. For each sample, prepare a 22 μl reaction mix (*see* **Note 6**).
   (a) <u>PCR mix for 1 reaction: EvaGreen assays</u>
   - 11 μl 2× ddPCR EvaGreen Supermix (BioRad)
   - 1.1 μl Forward primer (2 μM)
   - 1.1 μl Reverse primer (2 μM)
   - 4.8 μl H_2O (*see* **Note 7**)

   (b) <u>PCR mix for 1 reaction: Probe assays</u>
   - 11 μl 2× ddPCR Supermix for Probes (no dUTP, Bio-Rad)
   - 1.1 μl 20× probe assay FAM
   - 1.1 μl 20× probe assay HEX
   - 4.8 μl H2O (see Note 7)
2. Pipette 18 μl l of the mix into the wells of a Twin-Tec 96-well plate.
3. Add 4 μl l DNA sample and mix well by pipetting up and down multiple times.

3.2 Generate Droplets Using the QX200 Automated Droplet Generator

1. Cover the plate with a seal and spin briefly in a centrifuge.
2. Continue with droplet generation following the user manual for the Automated Droplet Generator.

3. The reaction volume after addition of the oil will be 70 µl. Emulsion PCR reactions are collected in a second 96-well plate.
4. Cover the plate with an aluminum foil and heat seal (*see* **Note 8**).

3.3 Perform Emulsion PCR in a Thermal Cycler

1. PCR can be performed in T100 thermal cycler (Bio-Rad) or an equivalent machine with programmable ramp rates.
   (a) Cycling profile for amplicons 50–250 bp with EvaGreen
      - 5 min 95 °C
      - 40 cycles—30 s 95 °C
         - 1-min annealing temperature
         - 5 min 4 °C
         - 5 min 90 °C
      - Hold at 12 °C

      Ramp rate for all steps is 2 °C/s
   (b) Cycling profile for amplicons 500–1500 bp with EvaGreen
      - 5 min 95 °C (ramp rate at 2 °C/s)
      - 40 cycles—30 s 95 °C (ramp rate at 2 °C/s)
         - 1 min annealing temperature (ramp rate at 2 °C/s)
         - 2 min 72 °C (ramp rate at 2 °C/s)
         - 5 min 4 °C (ramp rate at 1 °C/s)
         - 5 min 90 °C (ramp rate at 1 °C/s)
      - Hold at 12 °C (ramp rate at 1 °C/s)
   (c) Cycling profile for probe assays (ramp rate for all steps is 2 °C/s)
      - 10 min 95 °C
      - 40 cycles—30 s 94 °C
         - 1 min annealing temperature (60 °C for Bio-Rad assays)
         - 10 min 98 °C
      - Hold at 12 °C (*see* **Note 9**)

3.4 Count Positive Droplets with the QX200 Droplet Reader

1. After the PCR has finished, place the plate into the Droplet Reader holder.
2. Enter the experimental information into the system using the QuantaSoft™ software, and start the run.

3.5 Analyze Results

In the 96-well plate layout, the appropriate wells are selected and analysis is started with QuantaSoft. Data is visualized as 1-D (Fig. 3) or 2-D (Fig. 4) plots. Suboptimal results due to, e.g., non-specific amplification can be clearly identified (Fig. 5).

Threshold settings to differentiate between negative and positive droplets can be applied automatically or manually afterwards.

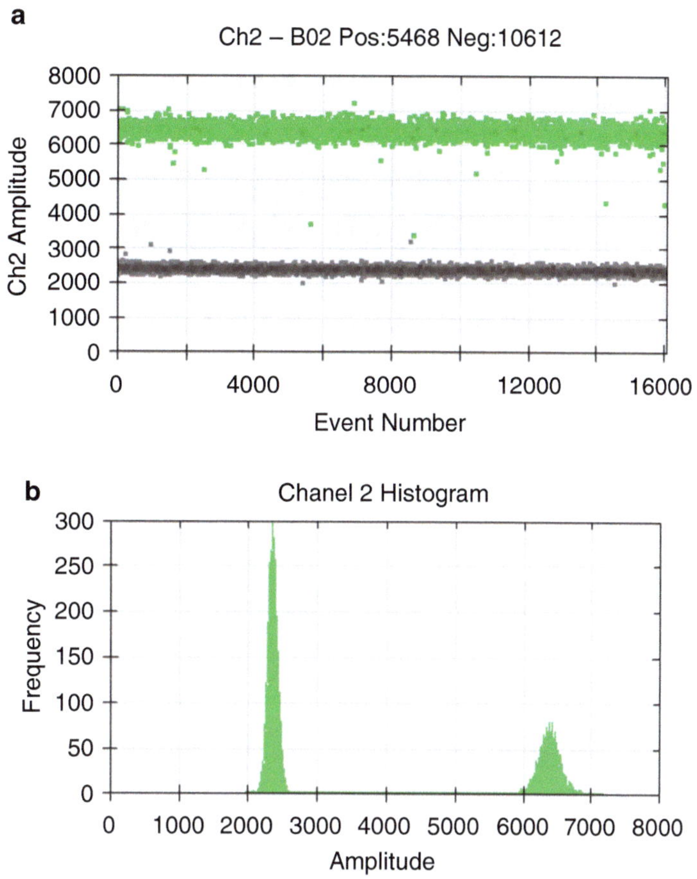

Fig. 3 1-D plot showing amplification of RPP30. A well-defined separation between negative droplets ((a) *lower line*, (b) *left peak*) and positive droplets ((a) *upper line*, (b) *right peak*) is seen, indicative of a well-optimized PCR reaction

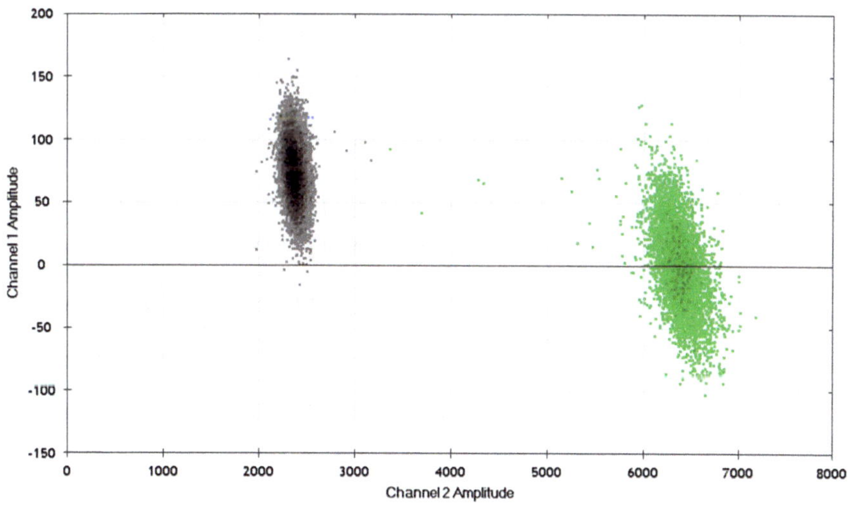

Fig. 4 2-D plot of the same reaction as in Fig. 3 showing negative droplets (*left cluster*) and positive events (*right cluster*)

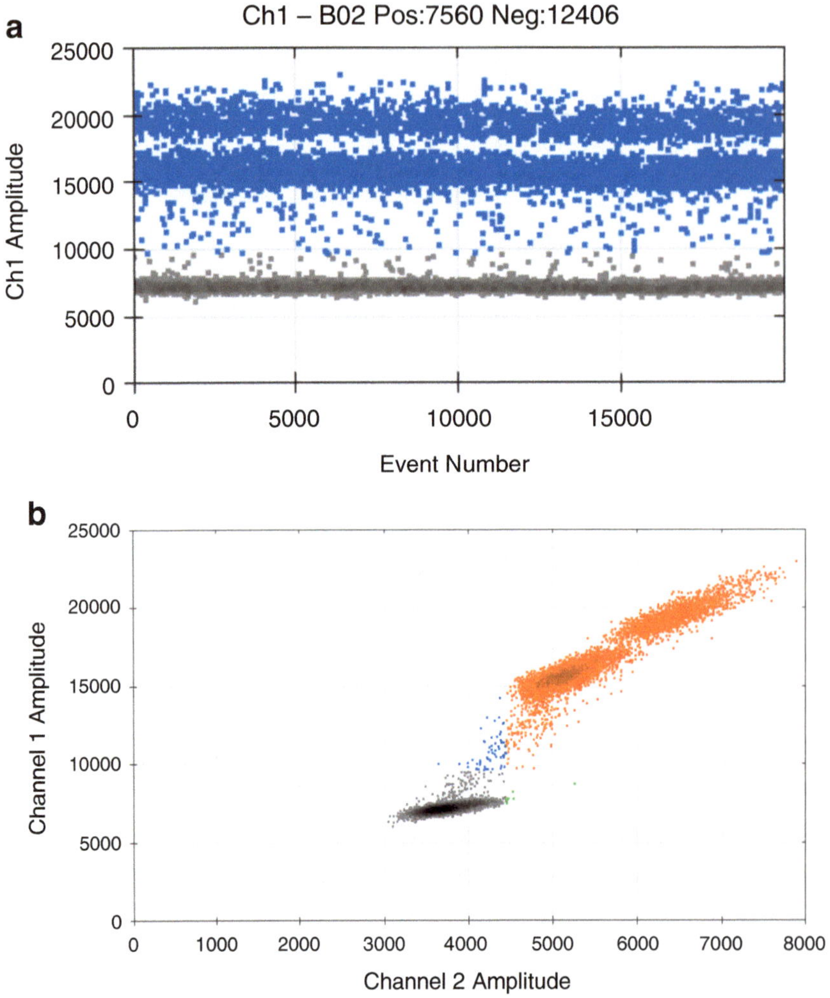

Fig. 5 1-D plot (*a*) and 2-D plot (*b*) of a suboptimal amplification showing double-positive droplet populations, indicative of by-product formation

The software calculates the copies/μl for each sample, using the ratio of positive to negative reactions. The calculation is based on a Poisson distribution ($x = \ln(1-p)$), where x is the number of DNA molecules in the reaction, and p is the proportion of positive reads.

For copy number calculation, the number of DNA molecules for the test locus should be divided by the number of molecules for the reference locus (usually assumed to be two for a diploid locus).

4 Notes

1. If the DNA being analyzed was isolated from cells that were replicating, then early-replicating regions of the DNA will be over-represented compared to late-replicating regions [6].

In such situations a control probe near the locus being analyzed is recommended [7].

2. Commonly used dye combinations are FAM+VIC or FAM+HEX.
3. Dilutions can best be made in a solution containing 10 mM Tris–HCl pH 8 and 0.05% Tween. Unpredictable results have been observed with low DNA concentrations diluted in pure H_2O.
4. Enzymes that have shown to work well in combination with the ddPCR chemistry are HaeIII (4-base cutter) and HindIII (6-base cutter), but other enzymes may also be suitable.
5. For any application, it is recommended that the "Minimum Information for Publication of Quantitative Digital PCR Experiments" guidelines are considered [8].
6. According to the original Bio-Rad manual, this reaction volume is 20 µl. Since the Automated Droplet Generator will take 20 µl for droplet generation, the reaction volume is set at 22 µl to ensure accurate transfer of the correct volume.
7. For CNV analysis add 2–5 U of the appropriate restriction enzyme (adjust H_2O volume accordingly), and incubate for at least 5 min at room temperature after the addition of the DNA.
8. Do not let a plate with emulsion reactions sit uncovered for more than 30 min. When the PCR is not performed directly after droplet generation, the plate can be stored at 4 °C for up to 4 h.
9. If not immediately measured, EvaGreen PCR reactions can be stored at 4 °C for up to 24 h. Probe reactions can be stored for up to 48 h at 4 °C.

References

1. Vogelstein B, Kinzler KW (1999) Digital PCR. Proc Natl Acad Sci U S A 96:9236–9241
2. Sykes PJ, Neoh SH, Brisco MJ et al (1992) Quantitation of targets for PCR by use of limiting dilution. Biotechniques 13:444–449
3. Miotke L, Lau BT, Rumma RT et al (2014) High sensitivity detection and quantitation of DNA copy number and single nucleotide variants with single color droplet digital PCR. Anal Chem 86:2618–2624
4. Bizouarn F (2014) Clinical applications using digital PCR. Methods Mol Biol 1160:189–214
5. Hindson BJ, Ness KD, Masquelier DA et al (2011) High-throughput droplet digital PCR system for absolute quantitation of DNA copy number. Anal Chem 83:8604–8610
6. Koren A, Polak P, Nemesh J et al (2012) Differential relationship of DNA replication timing to different forms of human mutation and variation. Am J Hum Genet 91:1033–1040
7. Usher CL, Handsaker RE, Esko T et al (2015) Structural forms of the human amylase locus and their relationships to SNPs, haplotypes and obesity. Nat Genet 47:921–925
8. Huggett JF, Foy CA, Benes V et al (2013) The digital MIQE guidelines: minimum information for publication of quantitative digital PCR experiments. Clin Chem 59:892–902

Chapter 12

Full-Length Mitochondrial-DNA Sequencing on the PacBio RSII

Rolf H.A.M. Vossen and Henk P.J. Buermans

Abstract

Conventional mitochondrial-DNA (MT DNA) sequencing approaches use Sanger sequencing of 20–40 partially overlapping PCR fragments per individual, which is a time- and resource-consuming process. We have developed a high-throughput, accurate, fast, and cost-effective human MT DNA sequencing approach. In this setup we first generate long-range PCR products for two partially overlapping 7.7 and 9.2 kb MT DNA-specific amplicons, add sample-specific barcodes, and sequence these on the PacBio RSII system to obtain full-length MT DNA sequences for genotyping/haplotyping purposes.

Key words Long-range PCR, Long read sequencing, PacBio RSII, Sample multiplexing, Haplotyping, MT-DNA

1 Introduction

Since the development of the polymerase chain reaction (PCR) technology by Kary Mullis in 1983 [1], it has become a widely used, versatile laboratory tool for use in genomics and transcriptomics research. With improvements made to increase amplicon lengths up to 2 kb as early as 1988 [2] and 42 kb in 1994 [3], long-range PCR (LR-PCR) technology is particularly useful to gain insight into complex genomic regions. Specific advantages over classical, short-amplicon, PCR include the ability to amplify large regions in a single reaction without the need for many individual reactions to cover the same region, and preserve the genomic context of the amplified region in one fragment, and the ability to exclude homologous regions by more specific primer design and to span repetitive sequences.

Recently, LR-PCR in combination with PacBio long read sequencing has been implemented for variant profiling using amplicons in the range of 2–5 kb, including the use of sample-multiplex barcodes via ligation of barcoded SMRTbell adapters [4] or by PCR [5, 6]. These studies illustrated the added benefits of PacBio long

Stefan J. White and Stuart Cantsilieris (eds.), *Genotyping: Methods and Protocols*, Methods in Molecular Biology, vol. 1492, DOI 10.1007/978-1-4939-6442-0_12, read sequencing in combination with LR-PCR, including the ability to obtain accurate phasing information or haplotyping of variants, high-accuracy consensus reads, opportunities for identifying rare variants with high confidence, an improved evenness of coverage compared to other target enrichment technologies (e.g., in solution capture), and the ability to exclude homologous regions by more specific primer design.

We have established a method for human MT DNA sequencing, using 7.7 and 9.2 kb LR PCR fragments in combination with PacBio sequencing, which results in high-quality full-length MT DNA sequences.

2 Materials

1. LA PCR Kit, Version 2.1 [Clontech RR013A].
2. SMRTbell Template Prep Kit 1.0 [Pacific Biosciences].
3. AMPure PB beads [Pacific Biosciences] or Washed Agencourt AMPure XP [Beckman].
4. DNA/Polymerase Binding Kit P6 V2 [Pacific Biosciences].
5. DNA Sequencing Reagent 4.0 [Pacific Biosciences].
6. MagBead Kit [Pacific Biosciences].
7. SMRT Cell 8Pac V3 [Pacific Biosciences].
8. Bioanalyzer DNA 12000 chip + reagents [Agilent].
9. Agarose and gel electrophoresis equipment.
10. 1 kb Ladder [Thermo Fisher].
11. 10 mM Tris–HCl, pH 8.5.
12. Qubit dsDNA HS Assay Kit and Qubit 2.0 Fluorometer [Thermo Fisher Scientific].
13. Locus-specific primers based on [7].

MT-DNA-specific primers with 5′-M13 tails: Standard desalted oligos (*see* **Note 1**).

Oligo name and Sequence 5′ → 3′

MT_1-2_M13F (9.2 kb) TGTAAAACGACGGCCAGT | ACATAGCACATTACAGTCAAATCCCTTCTCGTCCC

MT_1-2_M13R CAGGAAACAGCTATGACC | ATTGCTAGGGTGGCGCTTCCAATTAGGTGC

MT_3_M13F (7.7 kb) TGTAAAACGACGGCCAGT | TCATTTTTATTGCCACAACTAACCTCCTCGGACTC

MT_3_M13R CAGGAAACAGCTATGACC | CGTGATGTCTTATTTAAGGGGAACGTGTGGGCTAT

Generic barcoding primers with 3′-M13 sequences: HPLC-purified oligos

Oligo name and Sequence 5′ → 3′

PB_M13F_x_F GGTAG | 16 nt barcode | TGTAA AACGACGGCCAGT

PB_M13R_x_R GGTAG | 16 nt barcode | CAGGA AACAGCTATGACC

The 16 nt barcode sequences can be obtained from the PacBio website: https://github.com/PacificBiosciences/Bioinformatics-Training/blob/master/barcoding/pacbio_384_barcodes.fasta

The 16 nt barcode in the reverse primer is the reverse complement sequence of the barcode in the forward primer.

3 Methods

3.1 LR-PCR #1

1. Perform a separate amplification reaction for the 7.7 and 9.2 kb fragments. In a PCR tube combine the following reagents on ice (final concentrations) with 50–100 ng genomic DNA, 400 nM gene-specific M13 tailed primers, 400 μM of each dNTP, 1× PCR buffer with 2.5 mM $MgCl_2$, and 1 U Takara LA Taq. Total volume should be 25 μL. Mix gently and proceed to the LR-PCR.
2. Cycle parameters: Denaturation for 3 min at 95 °C, 30 cycles of two-step PCR: denaturation for 10 s at 98 °C and annealing and extension for 15 min at 68 °C. Final extension of 15 min at 68 °C.
3. Confirm the specificity of the products on a 1 % agarose gel or Bioanalyzer 12000 chip (Fig. 1a) (*see* **Note 2**).

3.2 LR-PCR #2: Barcoding Step

4. Measure the products on the Qubit, and pool equimolar amounts of the 7.7 and 9.2 kb fragments per individual.
5. Purify the fragments with 0.5× volumes AMPure PB beads as described in the "SMRTbell Template Prep Kit 1.0" kit protocol, and elute in 25 μL 10 mM Tris–HCl, pH 8.5.
6. Perform the barcode PCR separately on the pooled fragments from each individual, using a unique barcode primer set. In a PCR tube combine the following reagents on ice (final concentrations) with 3 μL of the Amplicon pool from LR-PCR #1, 400 nM barcoded M13 forward and reverse primers, 400 μM of each dNTP, 1× PCR buffer with 2.5 mM $MgCl_2$, and 1 U Takara LA Taq. Total volume should be 25 μL. Mix gently and proceed to the LR-PCR.
7. Cycle parameters: Denaturation for 3 min at 95 °C, 5 cycles: denaturation for 10 s at 98 °C and annealing and extension for 15 min at 68 °C. Final extension of 15 min at 68 °C.
8. Confirm the specificity of the products on a 1 % agarose gel or Bioanalyzer 12000 chip (Fig. 1b).

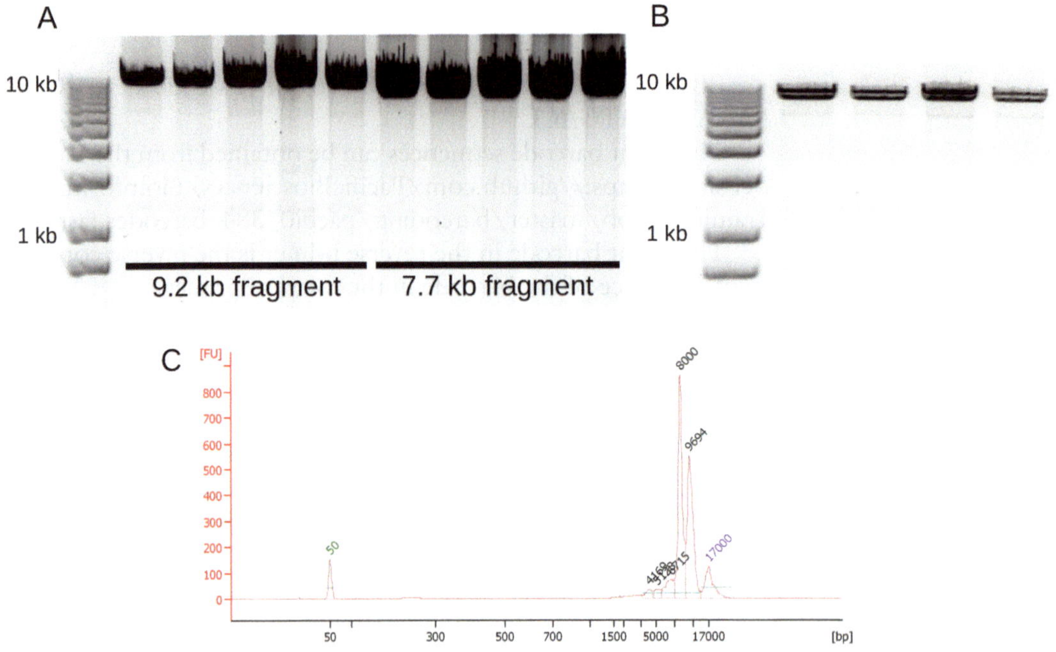

Fig. 1 LR-PCR products on 1 % TBE agarose gel after PCR #1 to amplify the separate 7.7 and 9.2 kb LR-PCR products (**a**), after pooling the separate LR-PCR products per individual and PCR #2 to introduce the sample-specific barcode (**b**). Bioanalyzer DNA 12000 chip trace (**c**) of the final library after pooling of all barcoded individuals and ligation of the PacBio SMRTbell adapter. The library fragments now include the primers and PacBio SMRTbell adapter sequences increasing the lengths to ~8.0 and 9.7 kb

3.3 PacBio SMRTbell Library Prep and Sequencing

9. Measure the products on the Qubit and make a mix containing equimolar, barcoded fragments of all individuals. Purify the fragments with 0.5× volumes AMPure PB beads (as for 5) and elute in 35 μL 10 mM Tris–HCl, pH 8.5.
10. Check the library on a Bioanalyzer 12000 chip and determine the concentration with the Qubit.
11. Proceed with the PacBio SMRTbell library preparation protocol starting from 1 to 5 μg of the amplified fragment pool following the 10 kb template preparation guidelines (Fig. 1c). Prepare the template binding complex and proceed to sequence the library with the P6/C4 chemistry with >3-h movie time.

3.4 Data Processing and Variant Calling

12. Run the long amplicon analysis (v1) protocol in the PacBio SMRT portal (v2.3.0) with the following analysis settings: min sub-read length = 7000; max number of sub-reads = 1000; ignore primer sequence = 35; trim ends = 35; only most supported = 0; cluster per gene fam = y; phase alleles = y; split results = n. Two high-quality haplogroup sequences of ~7.9 and

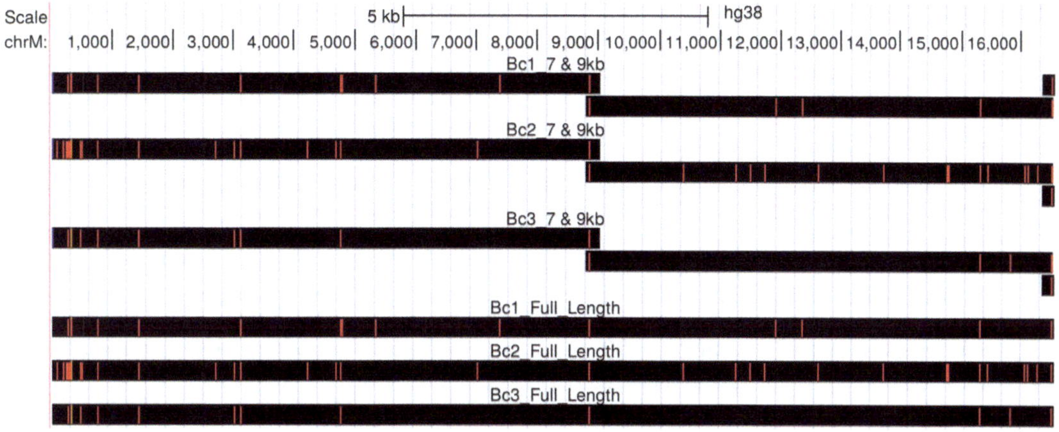

Fig. 2 Image from the UCSC genome browser displaying the PacBio read alignments (*black*) for the partially overlapping 7.7 and 9.2 kb amplicons (*top*) and the merged 16 kb full-length MT DNA sequences (*bottom*). The *red bars* indicate variants relative to the rCRS reference genome sequence

9.4 kb (this length includes the primers and PacBio SMRTbell adapter sequences) per individual should be produced (*see* **Note 3**). Remove the primer sequences from the reads.

13. Optional: Use CAP3 [8] (website: http://doua.prabi.fr/software/cap3) to merge the two partially overlapping sequences per individual into one full-length MTDNA sequence (*see* **Note 4**).
14. Align the MT-DNA sequences to the Revised Cambridge Reference Sequence (rCRS) of the Human Mitochondrial DNA sequence (NC_012920) with BWA MEM (v1.7.1), make bam and pileup files (Samtools v1.2), and determine the variants with bcftools (v1.2; bcftools call -mv -Ov -P 0.99 -p 0.99 | bcftools norm -m -both). Merge the vcf files from the two haplogroup sequences per individual into a single file (Fig. 2).

4 Notes

1. The human MT DNA-specific sequences can easily be substituted for sequencing of nonhuman MT DNA genomes.
2. This step is optional. When processing many samples in parallel, this step may be skipped in the interest of time.
3. When haplogroup sequences are found with lengths other than the expected ~7.9 and 9.4 kb, adjust the "max number of sub-reads" parameter and rerun the analysis.
4. Make sure that the two haplogroup sequences are in the same orientation.

References

1. Bartlett JS, Stirling D (2003) A short history of the polymerase chain reaction. Methods Mol Biol 226:3–6
2. Saiki RK, Gelfand DH, Stoffel S et al (1988) Primer-directed enzymatic amplification of DNA with a thermostable DNA polymerase. Science 239:487–491
3. Cheng S, Fockler C, Barnes WM, Higuchi R (1994) Effective amplification of long targets from cloned inserts and human genomic DNA. Proc Natl Acad Sci 91:5695–5699
4. Shukla SA, Rooney MS, Rajasagi M et al (2015) Comprehensive analysis of cancer-associated somatic mutations in class I HLA genes. Nat Biotechnol 33:1152–1158
5. Guo X, Lehner K, O'Connell K et al (2015) SMRT sequencing for parallel analysis of multiple targets and accurate SNP phasing. G3 Genes Genomes Genetics 5:2801–2808
6. Qiao W, Yang Y, Sebra R et al (2016) Long-read single molecule real-time full gene sequencing of cytochrome P450-2D6. Hum Mutat 37:315–323. doi:10.1002/humu.22936
7. Maitra A, Cohen Y, Gillespie SED et al (2004) The human MitoChip: a high-throughput sequencing microarray for mitochondrial mutation detection. Genome Res 14:812–819
8. Huang X, Madan A (1999) CAP3: a DNA sequence assembly program. Genome Res 9:868–877

Chapter 13

Targeted Locus Amplification and Next-Generation Sequencing

Quint P. Hottentot, M. van Min, E. Splinter, and Stefan J. White

Abstract

Despite developments in targeted and whole-genome sequencing, the robust detection of all genetic variation, including structural variants, in and around genes of interest and in an allele-specific manner remains a challenge. Targeted locus amplification (TLA) is a cross-linking-based technique that generates complex DNA libraries covering >100 kb of contiguous sequence surrounding one primer pair complementary to a short locus-specific sequence. In combination with next-generation sequencing, TLA enables the complete sequencing and haplotyping of targeted regions of interest. Here we outline the basis of TLA, together with a detailed protocol of the technique.

Key words Variant detection, Structural variation, Copy number variation, Next-generation sequencing

1 Introduction

Targeted locus amplification (TLA) [1] is based on cross-linking to connect DNA sequences that are in close physical proximity, followed by the fragmentation and religation of cross-linked DNA (Fig. 1). In the cited protocol a digestion is performed with a 4 bp restriction enzyme, the cross-linked DNA fragments are ligated, and a reverse cross-linking step is performed. A subsequent digestion with a 5 bp restriction enzyme (with a recognition site overlapping that of the 4 bp restriction enzyme used in the first step) followed by ligation generates circles of DNA. PCR amplification uses inverse primers that are located close to the restriction sites that define the primary "locus-specific sequence." The resulting ~2 kb products are then randomly sheared, and prepared for sequencing. Broad coverage of mapped DNA sequence allows variant detection over 100 kb of contiguous DNA sequence using a single pair of PCR primers, meaning for many genes exons and introns can be analyzed in a single reaction. TLA amplifications can be multiplexed across larger loci and/or multiple genes.

Stefan J. White and Stuart Cantsilieris (eds.), *Genotyping: Methods and Protocols*, Methods in Molecular Biology, vol. 1492,
DOI 10.1007/978-1-4939-6442-0_13, © Springer Science+Business Media New York 2017

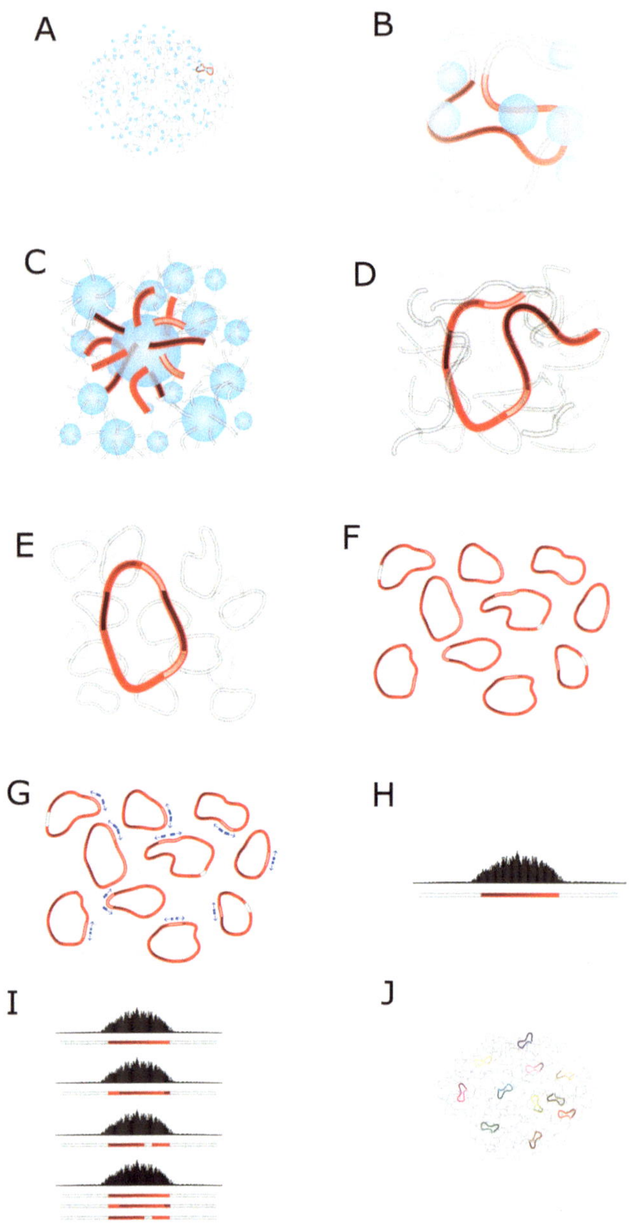

Fig. 1 Targeted locus amplification (TLA). (**a**) First, genomic DNA is cross-linked. (**b**) Cross-linking preferentially occurs between sequences in extreme physical proximity. This step therefore results in the cross-linking of sequences from the same locus (depicted in red). (**c**) The cross-linked DNA is fragmented, religated with a ligase enzyme, and then de-cross-linked. (**d**) This results in TLA template; long stretches of DNA consisting of religated DNA fragments originating from the same locus. (**e**) This template is fragmented and circularized. (**f**) Stochastic variation in the folding, cross-linking, and religation of DNA fragments in individual copies of a locus results in a repertoire of DNA circles that are composed of unique combinations of DNA fragments from that locus. (**g**) Circular fragments originating from the locus of interest are amplified with inverse primers complementary to a short locus-specific sequence. (**h**) As a result, the complete locus is amplified and can be sequenced using next-generation sequencing technologies. (**i**) In this manner the TLA technology enables targeted hypothesis-neutral sequencing. It detects all sequence and structural variants in loci of interest, also in heterogeneous samples such as tumors. (**j**) The TLA technology permits multiplexing. Multiple loci can be amplified in multiplex and/or multiple individual amplifications

TLA results in amplicons consisting of combinations of DNA fragments originating from the same individual copy of a locus. In combination with paired-end NGS sequencing and/or long read sequencing technologies TLA enables the haplotyping of regions of interest [2].

2 Materials

2.1 Equipment

1. Magnetic rack for Eppendorf tubes.
2. Microcentrifuge.
3. Centrifuge with swing-out rotor.
4. Orbital shaker.

2.2 Consumables

1. RBC lysis buffer: Weigh 4.13 g ammonium chloride and 0.5 g potassium bicarbonate and add 193.5 μl 0.5 M EDTA. Dissolve in 500 ml Milli-Q H_2O and filter sterilize (*see* **Note 1**).
2. Resuspension buffer: Add 1 ml fetal calf serum to 9 ml PBS (*see* **Note 2**).
3. Lysis buffer: 50 mM Tris-HCl (pH 7.5), 150 mM NaCl, 5 mM EDTA, 1% Triton X-100, 0.5% NP-40.
4. Formaldehyde: 37% Solution (*see* **Note 3**).
5. Methyl Green-Pyronin (*see* **Note 3**).
6. 1 M Glycine (*see* **Note 3**).
7. 1× Phosphate-buffered saline (PBS) pH 7.2: 137 mM NaCl, 2.7 mM KCl, 10 mM Na_2HPO_4, 1.8 mM KH_2PO_4 (*see* **Note 4**).
8. 5% SDS: Weigh 5 g SDS (*see* **Note 5**) in a 250 ml flask, add 85 ml Milli-Q H_2O, and mix with a magnetic stirrer. Heat to 68 °C if necessary and adjust the volume to 100 ml with Milli-Q H_2O. Mix again (*see* **Note 6**), and store at room temperature (*see* **Note 7**).
9. 2% TritonX-100: Mix 20 ml of Triton X-100 in 80 ml of Milli-Q H_2O. Store at 4 °C.
10. 10 mM Tris pH 7.5.
11. 10× Ligation buffer: 500 mM Tris–HCl, 100 mM $MgCl_2$, 10 mM ATP, 100 mM DTT, pH 7.5 at 25 °C. store at −20 °C (*see* **Note 8**).
12. NlaIII (New England Biolabs) (*see* **Note 9**).
13. NspI (New England Biolabs) (*see* **Note 9**).
14. 10× RE buffer 4 (New England Biolabs).
15. T4 DNA Ligase 5 U/μl (*see* **Note 9**).
16. Proteinase K 10 mg/ml (*see* **Note 9**).
17. RNase A 10 mg/ml (*see* **Note 9**).

18. Phenol-chloroform (*see* **Note 10**).
19. 3 M NaAc (pH 5.6).
20. 100% Ethanol.
21. 70% Ethanol.
22. Glycogen 20 mg/ml.
23. AMPure XP beads (*see* **Note 11**).
24. 5× PCR buffer (ThermoFisher).
25. 10 mM dNTPs.
26. Phire polymerase (ThermoFisher) (*see* **Note 9**).
27. ATP.
28. QIAquick PCR Purification Kit (optional, *see* **Note 12**).
29. PCR primers: The PCR amplification step uses inverse primers. These should be located close to the target sequence, as most reads will be located near the primers. There are several factors that need to be taken into account when designing the primers. They should be located between, and close to, two adjacent NlaIII restriction sites, directed outwards. The distance between the two sites should be at least 200 bp, to maximize the cross-linking efficiency with other DNA sequences. The primer sequences should be unique, and are not expected to give a product on genomic DNA. As for routine PCR, the primers should ideally have a GC content of 40–60%, be 18–22 nt in length, and have melting temperatures (T_m) that differ by <1 °C. They can be ordered as standard, desalted oligonucleotides.

3 Methods

3.1 White Blood Cell Isolation for TLA

The TLA protocol uses 5–10 million viable cells. This may require multiple 3 ml tubes of blood.

1. Mix the blood sample by inverting the tube ten times. Transfer 3 ml of blood to a 50 ml tube containing 25 ml RBC lysis buffer.
2. Mix five times by inversion. Incubate for 5 min at room temperature while agitating. The sample should become a clear red suspension.
3. Centrifuge for 10 min at 250×*g* in a swing-out centrifuge at room temperature. A 30–100 μl sized pink pellet should be observed. If the RBC lysis was incomplete then the pellet will be larger and dark red; **steps 3** and **4** should be repeated.
4. Pour off the supernatant and resuspend the pellet in 1 ml RBC lysis buffer. Transfer the sample to a labeled 1.5 ml reaction tube.
5. Centrifuge for 10 min at 250×*g* at room temperature. A 5–30 μl sized white pellet should be observed, covered by a thin layer of red cells.

6. Carefully pipette off the supernatant and red cells, ensuring that only the white cell pellet remains. Repeat **steps 5–8** if a pinkish/red pellet is observed.
7. Resuspend the pellet in 900 μl resuspension buffer and count the WBCs. Avoid platelets in this count, which are smaller than the WBCs.
8. The cells are now ready for the TLA protocol (Subheading 3.2), or can be stored at −80 °C (*see* **Note 13**).

3.2 TLA

3.2.1 Cross-Linking and Cell Lysis

1. Centrifuge 1×10^7 cells for 10 min at 250 × *g* at room temperature. Discard supernatant and resuspend the pellet in 1 ml resuspension buffer.
2. Add 56 μl 37% formaldehyde, mix well, and incubate for 10 min at room temperature (*see* **Note 14**).
3. Add 400 μl 1 M glycine and place tube immediately on ice.
4. Centrifuge for 2 min at 500 × *g*. Remove supernatant, resuspend pellet in 500 μl PBS, and centrifuge for 2 min at 500 × *g*.
5. Remove supernatant and resuspend pellet in 500 μl lysis buffer. Incubate for 5 min at room temperature, followed by 5 min at 65 °C and 1 min on ice.
6. Cell lysis control: Mix 3 μl of suspension with 3 μl of Methyl Green-Pyronin on a microscope slide, and cover with a cover slip. When viewed under a microscope, cytoplasm stains pink, and the nuclei/DNA stains blue/green.

3.2.2 Digestion

1. Centrifuge for 2 min at 1000 × *g*, and then remove supernatant.
2. Resuspend the pellet in 400 μl 1× RE buffer 4 and centrifuge for 2 min at 1000 × *g*.
   Remove all the supernatant and resuspend the pellet in 200 μl 1× RE buffer 4.
3. Place tube at 37 °C and add 12 μl 5% SDS. Incubate for 30 min at 37 °C while shaking at 900 rpm.
4. Add 30 μl 20% Triton X-100 and incubate for 30 min at 37 °C while shaking at 900 rpm. As the undigested control, take a 5 μl aliquot and add 90 μl 10 mM Tris pH 7.5.
5. Add 400 U NlaIII to the remainder, and incubate overnight at 37 °C while shaking at 900 rpm. As the digestion control 1, take a 5 μl aliquot of the sample and add to 90 μl 10 mM Tris pH 7.5.
6. Add 5 μl Proteinase K (10 mg/ml) to the undigested control and digestion control 1, and incubate for 1 h at 65 °C.
7. Load ~20 μl of each on a 0.6% agarose gel (*see* Fig. 2a). If the digestion control shows signs of incomplete digestion, repeat **step 5**.

3.2.3 Ligation and De-cross-linking

1. Inactivate the NlaIII enzyme in the remaining ~240 µl by incubating for 20 min at 65 °C, and then place tube for 1 min on ice.
2. Add a mix containing 210 µl milli-Q H_2O, 50 µl 10× ligation buffer, and 20 U T4 ligase and incubate for at least 2 h at room temperature.
3. Take a 10 µl aliquot of the sample as ligation control to determine the ligation efficiency. Add 80 µl 10 mM Tris pH 7.5 and 5 µl Prot K (10 mg/ml) and incubate for 1 h at 65 °C. Load ~20 µl on a 0.6% agarose gel (*see* Fig. 2b). If ligated product is observed, proceed with **step 4**. If ligated product is not observed, add fresh ATP and T4 ligase and repeat the 2-h incubation.
4. Add 5 µl Proteinase K (10 mg/ml) and de-cross-link overnight at 65 °C.

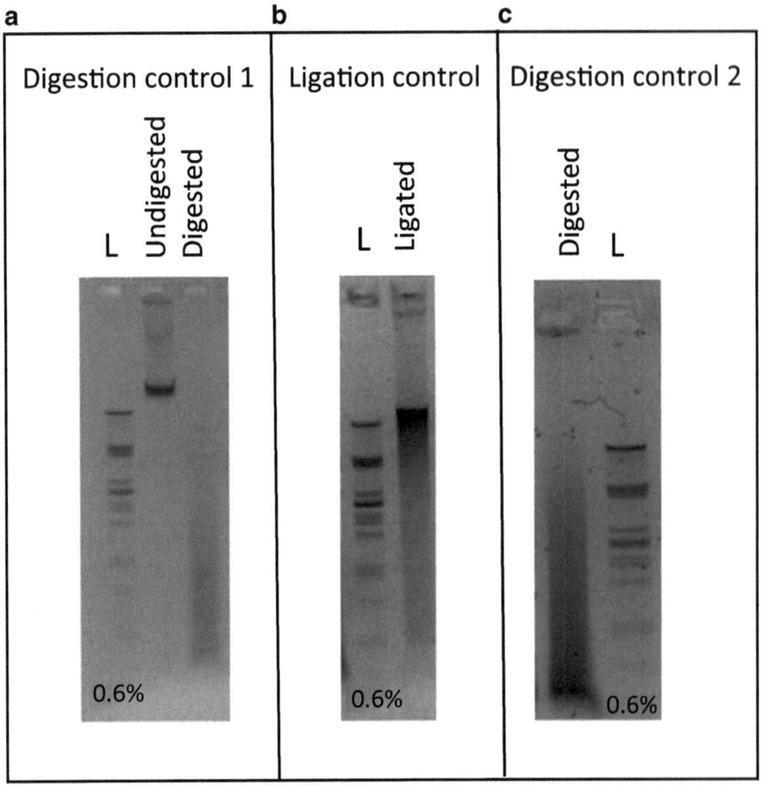

Fig. 2 Agarose gel electrophoresis of the different controls. (**a**) The NlaIII digestion should result in a very light smear between the 0.5 and 1.5 kb (digestion control 1). (**b**) The first ligation will result in the formation of DNA fragments of ~15 kb (ligation control). (**c**) The second digestion with NspI will result in a range of DNA fragments, averaging ~2 kb in length (digestion control 2). For each gel image L = DNA ladder, which is phage lambda digested with PstI

3.2.4 DNA Purification

1. Add 5 µl RNase A (10 mg/ml) and incubate for 10 min at 37 °C.
2. Add 500 µl phenol-chloroform and shake tube vigorously. Centrifuge for 4 min at 13,000 rpm in a microcentrifuge at room temperature.
3. Transfer 2× 250 µl of the aqueous phase to two new 1.5 ml tubes, and to each add:

   250 µl Milli-Q H_2O

   50 µl 2 M NaAC pH 5.6

   1 ml 100% Ethanol

   Mix thoroughly, and then place tubes at −80 °C until sample is frozen
4. Centrifuge for 20 min at 13,000 rpm in a microcentrifuge at 4 °C.
5. Remove supernatant and add 1 ml cold 70% ethanol. Centrifuge for 4 min at 13,000 rpm in a microcentrifuge at room temperature.
6. Remove the supernatant and briefly dry the pellet. Resuspend each pellet in 75 µl 10 mM Tris pH 7.5 at 37 °C, and combine the samples into one tube (*see* **Note 15**).
7. To the 150 µl sample add:

   50 µl 10× Restriction buffer

   5 µl NspI

   295 µl Milli-Q H2O

   Incubate overnight at 37 °C
8. As the digestion control 2, add a 5 µl aliquot of the sample to 95 µl 10 mM Tris pH 7.5. Run ~20 µl on a 0.6% agarose gel (Fig. 2c). If the sample is digested then continue to Subheading 3.2.5. If not, then repeat the digestion.

3.2.5 Ligation and Purification

1. Inactivate enzyme by incubating at 65 °C for 30 min.
2. Transfer sample to a 50 ml tube and add:

   1.4 ml 10× Ligation buffer

   12.1 ml Milli-Q H2O

   20 µl 5 U/µl T4 DNA ligase

   Ligate overnight at 16 °C
3. To the sample add:

   1.4 ml 2 M NaAC pH 5.6

   14 µl Glycogen (20 mg/ml)

   35 ml 100% Ethanol

Place at −80 °C until sample is frozen

4. Centrifuge for 45 min at 3200 × *g* at 4 °C.
5. Remove supernatant and add 15 ml cold 70% ethanol. Centrifuge for 15 min at 3200 × *g* at 20 °C.
6. Remove the supernatant and dry the pellet. Dissolve the pellet in 150 μl 10 mM Tris pH 7.5 at 37 °C and transfer the sample to a 2 ml Safe-Lock tube.
7. Add 270 μl AMPure XP beads and incubate for 5 min at room temperature.
8. Place the tube in a magnetic rack for at least 4 min, until the beads are separated to the side of the tube.
9. Carefully invert the magnetic rack four times to remove any liquid from the cap and incubate for 1 min. Remove the supernatant, making sure not to disturb the beads.
10. Remove the tube from the magnetic rack, add 500 μl freshly prepared 80% EtOH, and resuspend the beads.
11. Place tubes in the magnetic rack for at least 1 min, until the beads are separated to the side of the tube. Remove the supernatant, making sure not to disturb the beads.
12. Add 500 μl freshly prepared 80% EtOH, and leave undisturbed for at least 1 min.
13. Remove the EtOH and allow the pellet to dry (*see* **Note 16**).
14. Resuspend the beads in 155 μl 10 mM Tris pH 7.5, and incubate for 1 min.
15. Place tubes in the magnetic rack for at least 1 min, until the beads are separated to the side of the tube. Transfer 150 μl of the eluted DNA to a clean 1.5 ml Safe-Lock tube.
16. Measure the DNA concentration using the Qubit.

3.2.6 TLA PCR

1. Set up the TLA PCR reaction by preparing the following mix:

   800 ng TLA Template

   Milli-Q H2O to 142 μl

   40 μl 5× PCR buffer (ThermoFisher)

   10 μl 10 mM Primer mix

   4 μl 10 mM dNTPs

   4 μl Phire Polymerase (ThermoFisher)
2. Mix and divide evenly between four PCR tubes.
3. Run on the following PCR program.

   Step 1: 98 °C for 30 s

   Step 2: 98 °C for 5 s, 55 °C for 5 s, 72 °C for 2 min; repeat 33 times

Step 3: 72 °C 5 min

4. Pool the four 50 μl PCR reactions into a clean 1.5 ml Safe-Lock tube.
5. Add 300 μl AMPure XP Beads to the 200 μl PCR sample, mix five times by inversion, and incubate for 15 min while agitating.
6. Place the tube in a magnetic rack for at least 4 min, until the beads are separated to the side of the tube.
7. Carefully invert the magnetic rack four times to remove any liquid from the cap and incubate for 1 min. Remove the supernatant, making sure not to disturb the beads.
8. Remove the tube from the magnetic rack, add 900 μl freshly prepared 80% EtOH, and resuspend the beads.
9. Place the tube in a magnetic rack for at least 1 min, until the beads are separated to the side of the tube. Remove the supernatant.
10. Add 900 μl freshly prepared 80% EtOH, and leave undisturbed for at least 1 min.
11. Remove the EtOH and allow the pellet to dry (*see* **Note 16**).
12. Resuspend the beads in 105 μl 10 mM Tris pH 7.5, and incubate for 1 min.
13. Place tubes in the magnetic rack for at least 1 min, until the beads are separated to the side of the tube. Transfer 100 μl of the eluted PCR product to a clean 1.5 ml Safe-Lock tube.
14. Measure the concentration with the Qubit. The sample is ready for NGS library preparation.

3.3 Sequencing and Bioinformatic Analysis

Library preparation depends on the sequencing platform, but will usually involve a fragmentation step and the addition of linkers. We have successfully processed TLA products on different Illumina platforms, including MiSeq and HiSeq sequencers. The number of reads required depends on the amount of genomic sequence to be covered and the sensitivity required to detect variants. As an example, <1 million reads on a MiSeq were sufficient to provide 50-fold coverage at >98% of the ~81 kb *BRCA1* gene.

Bioinformatic analysis involves multiple steps. The first is alignment against a reference genome, which can be performed in two stages. We perform the initial alignment with BWA [3]. Reads that do not align during this step can undergo an in silico NlaIII digestion, with the resulting sequences aligned again. The two BAM files can then be combined for further analysis (Fig. *3*).

Many different approaches have been described for calling SNVs (reviewed in [4]), with GATK [5] and SAMtools [6] being

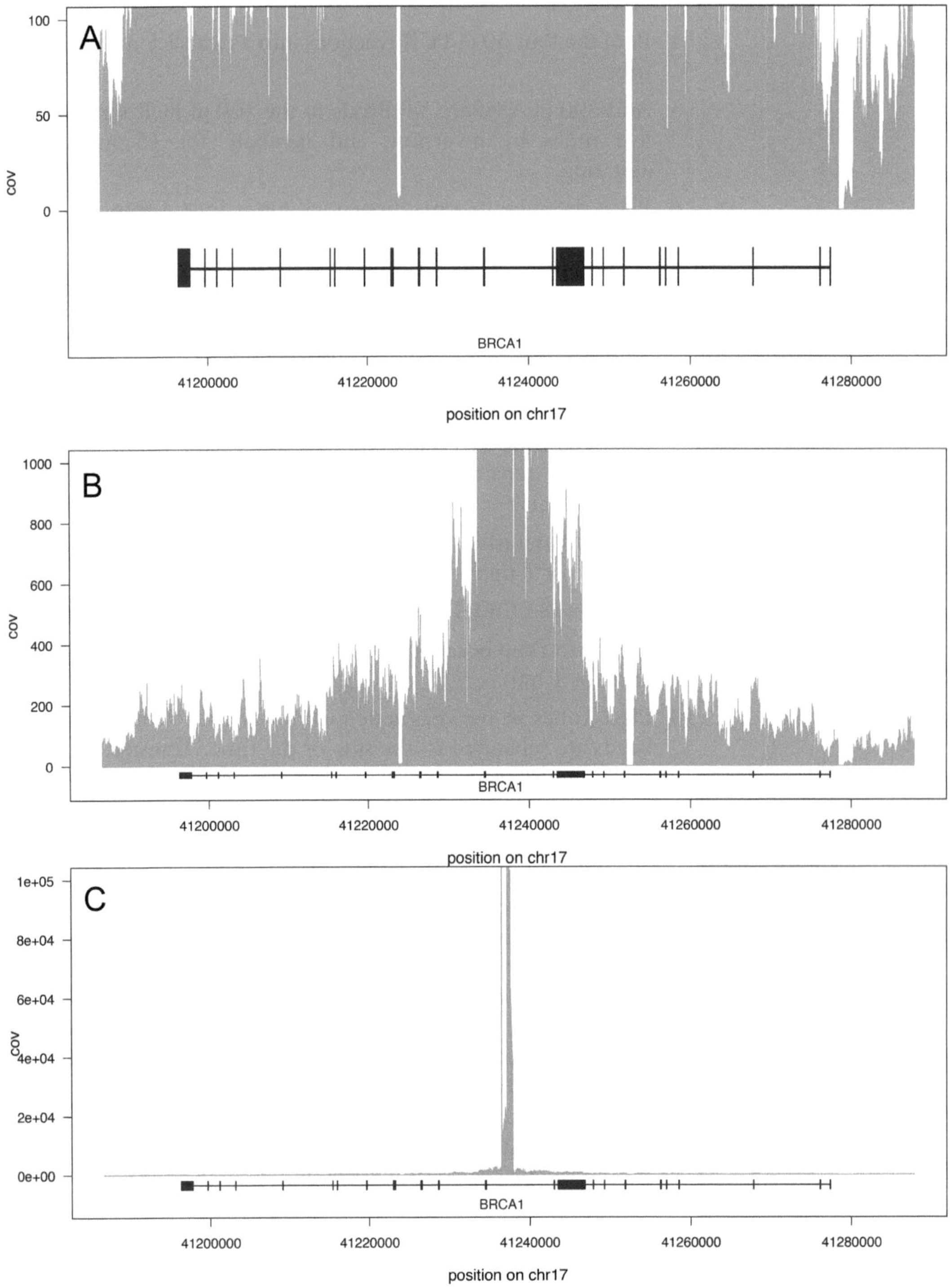

Fig. 3 Results of a TLA experiment, using a single primer pair within the *BRCA1* gene. Coverage profiles were generated from 450,000 paired-end, 300 bp reads. Different thresholds are indicated: (**a**) 100×, (**b**) 1000×, (**c**) 100,000×

popular options. The criteria used for determining a real variant will depend on the sample being analyzed. For example, a tumor-derived DNA sample may contain contaminating non-tumor DNA, so variants may not be present in 50% of the reads.

Structural variants can usually be detected initially by visual inspection of the aligned reads. Translocations will result in significant coverage on another chromosome. Deletions and inversions will result in a region of relatively increased coverage further away from the primer pair. SNV information in the intervening region can be used to distinguish between the two situations. A deletion should lead to a loss of heterozygosity (no heterozygous SNPs), which will not be observed with an inversion. The detection of breakpoint sequences will also clarify the type of rearrangement.

4 Notes

1. Store and use the RBC lysis buffer at room temperature.
2. This should be prepared fresh each time.
3. These can cause harm to the user and the environment when not used and disposed correctly. Safety instructions can be found in the MSDS.
4. Adjust the pH with HCl.
5. Wear a face mask and work in a fume hood to avoid inhalation of SDS particles.
6. Passing through a 0.45 μm filter will remove any undissolved material.
7. Do not store the solution at 4 °C, as SDS will precipitate at a temperatures below 15 °C. Should this happen, the SDS can be redissolved by warming in a water bath.
8. The ligation buffer is sensitive to multiple freeze and thaw cycles. Aliquot these reagents if more than three freeze and thaw cycles are expected.
9. Store at –20 °C, and keep on ice when in use.
10. Phenol-chloroform is harmful to the user and the environment when not used and disposed correctly. Safety instructions can be found in the MSDS.
11. Thoroughly vortex the AMPure XP beads before use.
12. The QiaQuick kit is an alternative to AMPure XP bead purification. It is faster and cheaper, but the yield will be lower.
13. The cells should be mixed thoroughly, but gently, with 100 μl DMSO in a cryovial. The tubes should be placed into a cryo-container and frozen at 1 °C/min.
14. Incubate for exactly 10 min. An advantage of using formaldehyde as a cross-linker is that the short reaction time minimizes

the formation of nonspecific cross-links, and allows the fixation of transient interactions.

15. If you do not plan to proceed with the analysis the protocol can be safely stopped here. Store the sample at –20 °C.
16. Over-drying the pellet may result in a dramatic yield loss.

Acknowledgements

We thank Marieke Simonis for data analysis.

References

1. de Vree PJ, de Wit E, Yilmaz M et al (2014) Targeted sequencing by proximity ligation for comprehensive variant detection and local haplotyping. Nat Biotechnol 32:1019–1025
2. Snyder MW, Adey A, Kitzman JO et al (2015) Haplotype-resolved genome sequencing: experimental methods and applications. Nat Rev Genet 16:344–358
3. Li H, Durbin R (2009) Fast and accurate short read alignment with Burrows-Wheeler transform. Bioinformatics 25:1754–1760
4. Nielsen R, Paul JS, Albrechtsen A et al (2011) Genotype and SNP calling from next-generation sequencing data. Nat Rev Genet 12:443–451
5. DePristo MA, Banks E, Poplin R et al (2011) A framework for variation discovery and genotyping using next-generation DNA sequencing data. Nat Genet 43:491–498
6. Li H, Handsaker B, Wysoker A et al (2009) The sequence alignment/map format and SAMtools. Bioinformatics 25:2078–2079

Chapter 14

Efficient, Cost-Effective, High-Throughput, Multilocus Sequencing Typing (MLST) Method, NGMLST, and the Analytical Software Program MLSTEZ

Yuan Chen and John R. Perfect

Abstract

Multilocus sequence typing (MLST) has become the preferred method for genotyping many biological species. It can be used to identify major phylogenetic clades, molecular groups, or subpopulations of a species, as well as individual strains or clones. However, conventional MLST is costly and time consuming, which limits its power for genotyping large numbers of samples. Here, we describe a new MLST method that uses next-generation sequencing, a multiplexing protocol, and appropriate analytical software to provide accurate, rapid, and economical MLST genotyping of 96 or more isolates in a single assay.

Key words Multilocus sequence typing, Genotyping, Next-generation sequencing, Multiplex PCR

1 Introduction

Multilocus sequence typing (MLST) can target multiple genomic loci, and the results can be easily archived, shared, and compared among laboratories. It is therefore considered one of the most reliable and informative methods for molecular genotyping [1, 2], even in the age of next-generation whole-genome sequencing. Many microorganisms have been genotyped using this method, and there is increasing interest in the variation among isolates and within microbial populations, especially in studies of microbial evolution, pathogenesis, ecology, and microbiomes [3–7]. MLST genotyping is a powerful approach to delineate species and strains, but the current methodology is costly, time consuming, and laborious.

Thus, we developed a high-throughput next-generation sequencing approach, NGMLST, to accelerate automation of the current MLST method [8]. The amplicon library preparation consists of two rounds of PCR. First, we adapted multiplex PCR in the first round of PCR to amplify all target loci at one time, which greatly reduces the labor. Second, in the next round of PCR a

Stefan J. White and Stuart Cantsilieris (eds.), *Genotyping: Methods and Protocols*, Methods in Molecular Biology, vol. 1492, DOI 10.1007/978-1-4939-6442-0_14, © Springer Science+Business Media New York 2017

unique barcode is added to each sample. All the amplicons are sequenced by Pacific Biosciences (PacBio) circular consensus sequencing (CCS) technology, which can then generate single-molecule consensus reads of 1–2 kb in length at a relatively low price. The sequencing reads of each locus from different samples are identified by MLSTEZ using different barcode and locus primer sequences. To obtain high-quality genotyping data, consensus sequences of each locus are generated by MLSTEZ automatically.

2 Materials

1. Diluted genomic DNA for each sample with a concentration of approximately 2.5 ng/μL.
2. Locus-specific primers with a 20 nt universal primer at the 5′ end (Fig. 1) (*see* **Note 1**).
3. Barcode primer, including a 5 nt padding sequence (GGTAG) at the 5′ end, followed by the 16 nt barcode sequence as suggested by PacBio (http://www.smrtcommunity.com/servlet/servlet.FileDownload?file=00P7000000W067VEAR), and a 20 nt universal primer added to the 3′ end (*see* **Note 2**).
4. Agarose.
5. Multiplex PCR Plus Kit (QIAGEN).
6. LongAmp Taq DNA Polymerase (New England BioLabs Inc.).
7. QIAquick PCR Purification Kit (QIAGEN).
8. DNA Template Prep Kit 2.0 (Pacific Biosciences).
9. Thermocycler (BioRad).
10. NanoDrop (Thermo Scientific).
11. Electrophoresis tank and power supply.

3 Methods

Keep all PCR preparations on ice. Two rounds of PCR reactions are required in this method.

1. Prepare 25 μL multiplex PCR mixture: 1 μL diluted genomic DNA, primer pairs for different locus at the optimized concentration, 12.5 μL 2× Master Mix (QIAGEN Multiplex PCR Plus Kit) (*see* **Note 3**).
2. Thermocycling conditions for multiplex PCR:
   (a) 95 °C for 5 min
   (b) 35 cycles of 30 s at 95 °C, 1.5 min at 58 °C, and 1.5 min at 72 °C
   (c) 10 min at 68 °C

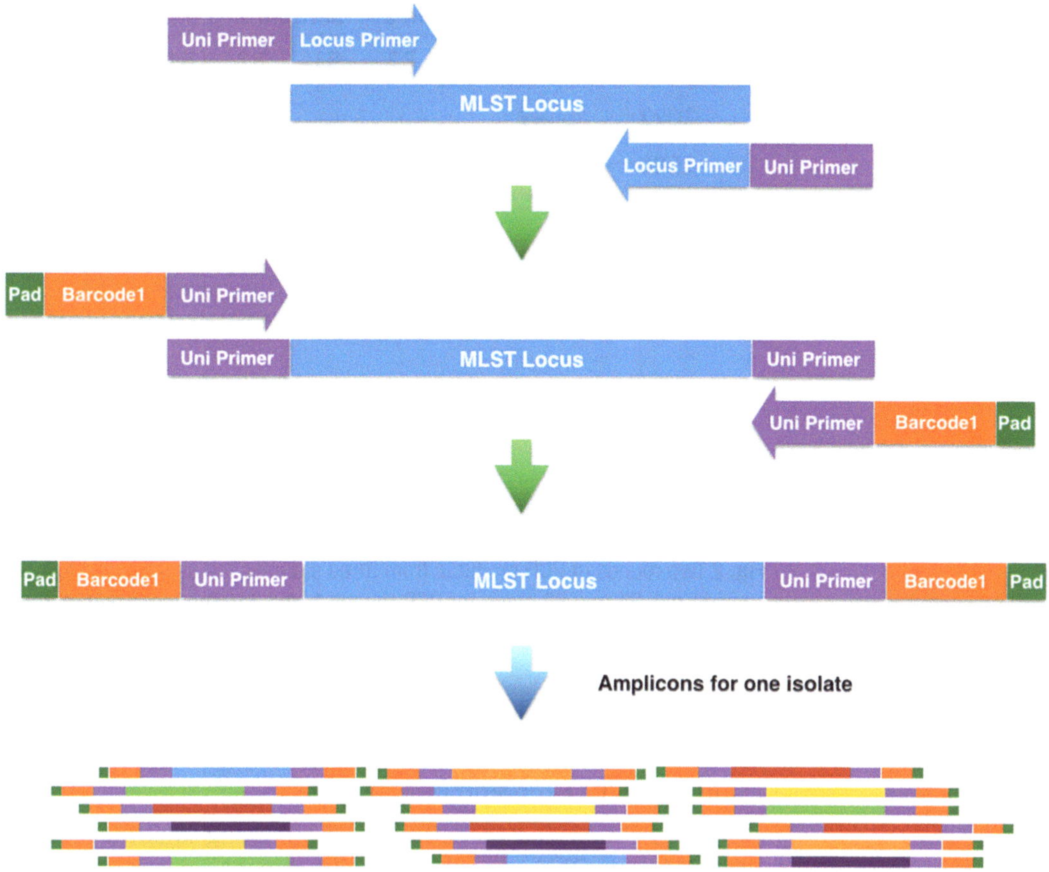

Fig. 1 Two rounds of PCRs are employed in NGMLST. In the first PCR round, each primer consists of a locus-specific sequence (*blue*, *see* **Note 1**) and a 20 nt universal primer sequence (*purple*). The barcode primers consist of three parts: (1) a 20 nt universal sequence (*purple*), which amplifies the template; (2) a 16 nt barcode sequence (*orange*) that identifies the amplicons from each different isolate; (3) and a 5 nt padding sequence (*green*) to provide equivalent binding affinities for adding the PacBio sequencing adapters. The final products of each isolate would have the same sequence structure on both ends, flanking different target locus sequences in the middle, which are shown with different colors (reproduced with permission from [8])

3. Visualize the multiplexed products on 1.2–2% agarose gel (optional step, gel concentration is based on product size), and make 1:50 dilutions of the multiplexed products.
4. Prepare 25 μL PCR mixture: 1 μL diluted multiplexed product, 5 μL 5× LongAmp Taq reaction buffer, 0.75 μL 10 mM dNTPs, 10 μL barcode primer for each.
5. Thermocycling conditions for second round of PCR:
   (a) 94 °C for 30 s
   (b) 35 cycles of 30 s at 94 °C, 30 s at 50 °C, and 60 s at 65 °C
   (c) 10 min at 65 °C

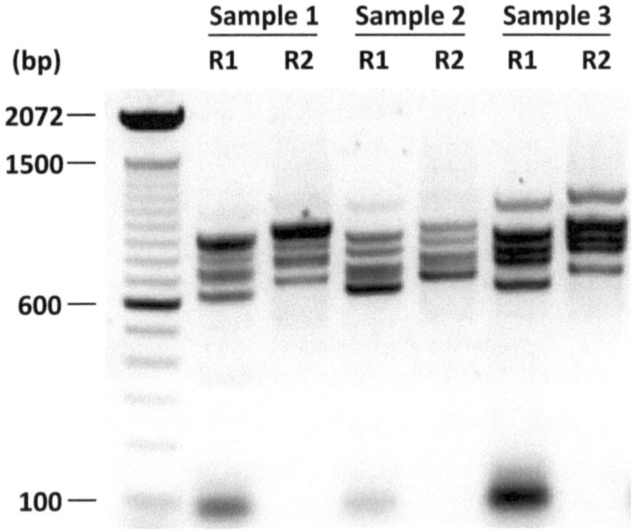

Fig. 2 Two rounds of PCR products from three samples are shown, on 1.4 % TAE agarose gel. R1 and R2 stand for PCR rounds 1 and 2 (reproduced with permission from [8])

6. Visualize the PCR products on 1.2–2 % agarose gel and estimate product concentration (Fig. 2).
7. Pool PCR products together based on having similar amount of DNA. The DNA concentration of each pool is determined using a Nanodrop ND-1000 Spectrophotometer. Multiple pools can be combined with equal amount of DNA (*see* **Note 4**).
8. Prepare SMRT Cell sequencing library using Pacific Biosciences DNA Template Prep Kit 2.0 according to the 3 or 10 kb template preparation and sequencing protocol. Instead of using magnetic beads, the amplicons are loaded by diffusion at a concentration of 300 pM. The pool is sequenced with a SMRT Cell on the PacBio RS II platform, using 1 × 180 min movie with P4-C2 chemistry (*see* **Note 5**).
9. Use PacBio SMRT Analysis to perform primary analysis, and the filtering parameters as follows: minimum polymerase read quality of 0.75; minimum read length of 50 bp; and minimum subread length of 50 bp. Filter out circular consensus sequencing (CCS) reads with fewer than four full passes (*see* **Note 6**).
10. Use the CCS FASTQ file as well as primer and barcode sequences as input for MSLTEZ (http://sourceforge.net/projects/mlstez/?source = directory) to generate consensus sequences for each locus and predict heterozygous loci (*see* **Note 7**).

4 Notes

1. The locus-specific primers used in previous studies can be adapted to NGMLST. Because of the length limitation of PacBio CCS method, the target locus length needs to be less than 2 kb, and the maximal difference in length between the amplicons cannot exceed 500 bp to avoid affecting the sequencing yield. The length limitation might change with the development of the sequencing platform. Other sequencing platforms can also be used for NGMLST, as long as the read length is longer than the lengths of the final amplicons. Two universal primers have been tested in the system: Primer 1, 5′-CTGGAGCACGAGGACACTGA; Primer 2, 5′-GCTGTCAACGATACGCTACG.
2. Ninety-six barcode sequences with Universal Primer 1 can be downloaded from the supplemental materials of ref. 8. Barcode sets can be designed symmetrically (same universal primer and barcode on either end) or asymmetrically (different universal primers and barcodes on either end), and both barcode sets are fully supported by the latest version of MLSTEZ.
3. The concentration of locus-specific primers needs to be optimized to obtain equal amounts of each product. The number of target MLST loci and tested isolates needs to be balanced. Based on the current throughput of PacBio CCS method, we suggest analyzing no more than 11 loci with lengths around 1 kb for 96 samples in one SMRT Cell.
4. For 96 samples, the amplicons can be pooled into four groups with 24 samples each for purification. For successful PCR amplification, 2 μL of each final product should be enough. The four purified pools are mixed together with equal concentration of DNA.
5. The loading concentration and movie time may vary depending on the version of DNA Template Prep Kit and chemicals used for sequencing.
6. The filter of minimal read length and subread length can be adjusted according to the target locus length. We suggest to use CCS reads with four or more passes. Sequencing reads with fewer passes usually have worse quality, which may affect the final genotyping result.
7. MLSTEZ can be used for analyzing sequences with FASTA or FASTQ format from any sequencing platform. However, we strongly recommend to use FASTQ format, which includes sequencing quality information. The default settings of MLSTEZ require at least three reads for each locus to generate consensus sequences. For PacBio CCS method, the output

files include CCS reads and subreads, and only CCS reads can be used for MLSTEZ. It is common to have several samples in one batch of an experiment that do not have enough reads to generate consensus sequences. These samples can be sequenced in additional batches, and all the sequencing reads of the same sample that are from different batches can be merged together using the "merge project" function of MLSTEZ. More details about MLSTZ are available online (http://sourceforge.net/p/mlstez/wiki/Manual/).

Acknowledgment

This work was supported by Public Health Service Grants AI73896 and AI93257 (JRP).

References

1. Schwartz DC, Cantor CR (1984) Separation of yeast chromosome-sized DNAs by pulsed field gradient gel electrophoresis. Cell 37:67–75
2. Maiden MC, Bygraves JA, Feil E et al (1998) Multilocus sequence typing: a portable approach to the identification of clones within populations of pathogenic microorganisms. Proc Natl Acad Sci 95:3140–3145
3. Litvintseva AP, Mitchell TG (2012) Population genetic analyses reveal the African origin and strain variation of Cryptococcus neoformans var grubii. PLoS Pathog 8:e1002495
4. Meyer W, Aanensen DM, Boekhout T et al (2009) Consensus multi-locus sequence typing scheme for *Cryptococcus neoformans* and *Cryptococcus gattii*. Med Mycol 47:561–570
5. Chen Y, Toffaletti DL, Tenor JL et al (2014) The *Cryptococcus neoformans* transcriptome at the site of human meningitis. mBio 5:e01087–13
6. Byrnes EJ, Bildfell RJ, Frank SA et al (2009) Molecular evidence that the range of the Vancouver Island outbreak of *Cryptococcus gattii* infection has expanded into the Pacific Northwest in the United States. J Infect Dis 199:1081–1086
7. Chen Y, Litvintseva AP, Frazzitta AE et al (2015) Comparative analyses of clinical and environmental populations of *Cryptococcus neoformans* in Botswana. Mol Ecol 24:3559–3571
8. Chen Y, Frazzitta AE, Litvintseva AP et al (2015) Next generation multilocus sequence typing (NGMLST) and the analytical software program MLSTEZ enable efficient, cost-effective, high-throughput, multilocus sequencing typing. Fungal Genet Biol 75C:64–71

Chapter 15

Rapid SNP Detection and Genotyping of Bacterial Pathogens by Pyrosequencing

Kingsley K. Amoako, Matthew C. Thomas, Timothy W. Janzen, and Noriko Goji

Abstract

Bacterial identification and typing are fixtures of microbiology laboratories and are vital aspects of our response mechanisms in the event of foodborne outbreaks and bioterrorist events. Whole genome sequencing (WGS) is leading the way in terms of expanding our ability to identify and characterize bacteria through the identification of subtle differences between genomes (e.g. single nucleotide polymorphisms (SNPs) and insertions/deletions). Modern high-throughput technologies such as pyrosequencing can facilitate the typing of bacteria by generating short-read sequence data of informative regions identified by WGS analyses, at a fraction of the cost of WGS. Thus, pyrosequencing systems remain a valuable asset in the laboratory today. Presented in this chapter are two methods developed in the Amoako laboratory that detail the identification and genotyping of bacterial pathogens. The first targets canonical single nucleotide polymorphisms (canSNPs) of evolutionary importance in *Bacillus anthracis*, the causative agent of Anthrax. The second assay detects Shiga-toxin (*stx*) genes, which are associated with virulence in *Escherichia coli* and *Shigella spp.*, and differentiates the subtypes of *stx*-1 and *stx*-2 based on SNP loci. These rapid methods provide end users with important information regarding virulence traits as well as the evolutionary and biogeographic origin of isolates.

Key words Pyrosequencing, Genotyping, SNP, canSNP, *Bacillus anthracis*, *Escherichia coli*, Shiga-toxin

1 Introduction

The differentiation of bacterial isolates remains a primary objective for microbiologists and is particularly important in clinical microbiology and food safety testing. Phenotypic methods such as colony morphology and biochemical characterization have become routine in microbiological testing; however, they are unable to differentiate at the subspecies level or perform outbreak tracking and thus may be missing important details. Recent advancements in molecular methods have revolutionized bacterial detection and strain typing, allowing unprecedented resolution of bacterial

Stefan J. White and Stuart Cantsilieris (eds.), *Genotyping: Methods and Protocols*, Methods in Molecular Biology, vol. 1492, DOI 10.1007/978-1-4939-6442-0_15, © Springer Science+Business Media New York 2017

genotypes through the use of sequencing to detect single nucleotide polymorphisms (SNPs), single nucleotide repeats (SNRs), as well as insertions and deletions (indels). Genotyping can be accomplished rapidly by pyrosequencing using the PyroMark Q24 system (Qiagen), a sequencing by synthesis platform which monitors the incorporation of nucleotides via the release of pyrophosphate [1]. Prior to pyrosequencing, polymerase chain reaction (PCR) products are generated using targeted PCR primers, one of which is biotinylated to allow the PCR product to be captured on streptavidin-coated sepharose beads. Pyrosequencing reads allow for rapid detection of SNPs, indels, and other sequence variants by generating real-time sequence information and can be used for detection and genotyping of bacterial pathogens [2–4].

SNPs are important markers for the typing and source tracking of bacterial pathogens, especially in outbreak scenarios [5–8]. SNP typing can be performed with high-throughput techniques (e.g. high resolution melt analysis and pyrosequencing), which allow for the rapid analysis of the large number of isolates typically associated with an outbreak [3, 9, 10]. With this in mind, we developed pyrosequencing assays for the PyroMark Q24 system for SNP typing of *Bacillus anthracis*, the etiologic agent of Anthrax. This bacterium is of both historical and present importance, as it has been developed as a biological weapon by several nations in the past and has gained recent notoriety due to its effective use as a bioterrorist agent (e.g. US Amerithrax incident of 2001 and the 1999 unsuccessful aerosol attack in Japan) [5–8, 11]. SNPs are of particular interest because of the inherent evolutionary stability present in the targets due to low mutation rates (10^{-10} per site per generation [12]) and the highly clonal nature of *B. anthracis* [13]. As part of its lifecycle *B. anthracis* forms highly resilient, quiescent spores which can persist in the environment for decades, contributing to a highly conserved genome [14]. This reduces the probability of unrelated strains sharing the same SNP allele (homoplasy) and ensures that the mutational changes observed are clade specific [5]. While there are a multitude of SNPs that can be identified using whole genome sequencing, these have been down selected to a core set of canonical SNPs (canSNPs) that can be used to identify the major lineages in the evolutionary history of *B. anthracis* [13]. Analyzing these slow evolving SNPs can identify clonal lineages and provide a tool for source attribution and outbreak tracking [14]. The pyrosequencing assays developed were designed to detect and characterize the canSNP profiles of *B. anthracis* in order to type and classify the isolates. These canSNP profiles were then used to produce a phylogenetic tree detailing the evolutionary relatedness between isolates, and can further be used for epidemiological analyses.

Pyrosequencing technology has recently been applied to the detection and subtyping of the Shiga-toxin genes in *Escherichia coli* and *Shigella spp.* [4]. Shiga-toxin producing *E. coli* (STEC) including O157:H7 are one of the most frequent sources of foodborne illness worldwide and the detection of Shiga-toxin genes is the first screening tool for detecting STEC in molecular diagnostic laboratories. Shiga-toxin (Stx) is the enterotoxin genes encoding stx are secreted by STEC and *Shigella spp.* that causes gastroenteritis and in some cases hemolytic-uremic syndrome (HUS) in humans; the genes encoding *stx* are known to have genotypic and phenotypic variation. According to the nomenclature by Scheutz et al. [16], there are 3 Shiga-toxin 1 gene subtypes (*stx*-1a, -1c, and -1d) and 7 Shiga-toxin 2 subtypes (*stx*-2a, -2b, -2c, -2d, -2e, -2f, and -2g). Shiga-toxin subtyping was traditionally performed by the reaction to antisera and Restriction Fragment Length Polymorphism (RFLP) analysis [17] and then shifted to single- or multiplex-PCR [16]. Shiga-toxin subtypes have been reported in relation to the pathogenicity of the human illness. Some Shiga-toxin subtypes, including *stx*-2a and *stx*-2c, are present in highly virulent STEC such as O157:H7 [17, 18]; in contrast *stx*-2e, *stx*-2f and *stx*-2g are present mostly in animal isolates and are rarely associated with human food-poisoning cases [19, 20]. The pyrosequencing application we introduce here is a rapid, cost-efficient, and robust assay for genotyping STEC. By further characterizing the *stx*-1a subtype by three SNP loci in the gene, we demonstrate that *stx*-1a type I and II are detected frequently from the Top-7 priority serotypes (O26, O45, O103, O111, O121, O145, and O157). These serotypes are most frequently implicated in severe clinical illness [4]. Thus, Shiga-toxin subtyping provides valuable information associated with food-borne human illness investigation and risk assessment for environmental and food samples.

The two methods detailed herein are for genotyping the bacterial pathogens *B. anthracis* and Shiga-toxin producing *E. coli* and *Shigella spp.* via pyrosequencing. The methods are rapid, highly specific, and generate sequence information in support of the positive identification and differentiation of isolates through SNP loci. Deploying such methods allows for adequate response measures in the event of an accidental or intentional food contamination.

2 Materials

2.1 Software

1. PyroMark Assay Design software.
2. PyroMark Q24 software.
3. Geneious R8 software (Biomatters).
4. NCBI BLAST tool (http://www.ncbi.nlm.nih.gov/blast/Blast.cgi) [21].
5. Literature search tools (e.g. Scopus).

2.2 Reagents

1. PyroMark PCR master mix (Qiagen) (*see* **Note 1**).

2. DNAse/RNase-Free water.
3. Oligonucleotide primers.
4. DNA template.
5. PyroMark Gold Q24 reagents (Qiagen).
6. 70% ethanol.
7. PyroMark binding buffer (Qiagen).
8. Streptavidin-coated Sepharose High Performance (HP) beads (GE Healthcare).
9. PyroMark Annealing buffer (Qiagen).
10. PyroMark Denaturation buffer (Qiagen).
11. PyroMark Wash buffer (Qiagen).

2.3 Disposables and Equipment

1. BSL-2 personal protective equipment (lab coat, eye protection, and disposable gloves).
2. PCR tubes.
3. Pipettor and filter tips.
4. Thermocycler.
5. PyroMark Q24 system (Qiagen).
6. Break-apart 96-well plates (VWR).
7. PyroMark Q24 plate (Qiagen).
8. PyroMark Q24 vacuum workstation (Qiagen).
9. PyroMark workstation troughs (Qiagen).
10. Heat block set at 80 °C.
11. PyroMark Q24 plate holder (Qiagen).
12. PyroMark Q24 Cartridge (Qiagen).
13. Orbit P2 Digital Shaker with 3 mm orbit size (Labnet).

3 Methods

3.1 Isolation of Genomic DNA

A variety of commercial DNA extraction methods can be used provided the DNA is sufficiently pure and free from endogenous proteins; higher quality DNA extractions can provide longer storage times. For the extraction of gram positive organisms (e.g. *B. anthracis*), including spores, the Epicenter MasterPure Gram positive kit is recommended as it has been previously demonstrated to be superior to other readily available kits [22]. For gram negative organisms, such as *E. coli*, a simple and rapid boiling lysis method may be used; however, more extensive extraction methods should also work well (*see* **Note 2**) [4].

3.2 Polymerase Chain Reaction (PCR)

PCR primers can be designed using a variety of software suites and we have used assays designed with both the PyroMark Assay Design software (Qiagen) as well as Geneious R8 (Biomatters). There are a few special primer/target considerations that must be made relative to normal PCR primers. The forward or reverse primer must have a 5′- biotin label so the PCR product can be immobilized on the streptavidin sepharose beads and the sequencing primer must be reverse-complementary to the strand containing the biotin linker. Sequencing targets containing homopolymers (greater than 4) should be avoided as they are challenging to resolve using pyrosequencing (*see* **Note 3**) [23]. Literature searches can help to quickly identify targets for a specific organism; otherwise whole genome alignments can be performed between the target organism and closely related species to identify unique regions. All primers and amplicons should go through a thorough BLAST search to ensure specificity.

1. Remove PyroMark PCR Master Mix components and PCR primers from −20 °C and thaw at room temperature.
2. In a template-free area (*see* **Note 4**), prepare master mix as per Table 1, vortex briefly and dispense into PCR tubes.
3. Move the PCR tubes containing master mix to template addition area and add ~5 ng of genomic DNA (see Table 1 for manufacturer's PCR reaction mix setup). Include a negative control, positive control, and internal control if available (*see* **Note 5**).
4. Seal PCR tubes and place in the thermocycler. Run suitable PCR program for amplification.

Table 1
Master mix for PCR

Reagent	Initial conc. (μM)	Final conc. (μM)	Volume, 1rxn (μL)
Ultrapure dH_2O			3
PyroMark Master	2×	1×	12.5
$MgCl_2$	25	0.5	0.5
Q solution	5×	1×	5
Forward primer	10	0.4	1
Reverse primer	10	0.4	1
Total volume			23
Sample DNA			2

5. Resolve PCR amplicons using the QIAxcel system or traditional electrophoretic gel to verify product presence and sizes prior to pyrosequencing (*see* **Note 6**).

3.3 Preparing a PyroMark Q24 Run File

1. Load the PyroMark Q24 software.
2. Click "File → New Assay" and select the appropriate assay (*see* **Note** 7).
3. In the text box, enter the desired dispensation (*see* **Note 8** and Table 2).
4. Save the dispensation file and create a new run (AQ or SQA).
5. Indicate the instrument method from the drop-down menu (*see* **Note 9**).
6. Set up the plate by selecting the plate locations and loading the appropriate assay from **step 2**.
7. Enter pertinent sample information.
8. Print the Protocol Setup for information on volumes needed for setting up the reagent cartridge (Tools → Show protocol setup).
9. Save the run file to a USB drive to be used later.

3.4 Pyrosequencing

1. Thaw sequencing primer and make a master mix containing 0.3 μM sequencing primer in a final volume of 25 μL annealing buffer (e.g. 0.75 μL of 10 μM primer and 24.25 μL buffer) for each sample in a microfuge tube. This tube will be used during the primer annealing step.
2. Make pyrosequencing plate.
3. Add PCR product such that the final volume of master mix and PCR product is 80 μL.
4. Set plate on an orbital shaker for 10 min at 4400 × *g* (*see* **Note 10**).
5. Take the tube prepared in **step 1** and add annealing buffer. Vortex briefly and dispense 25 μL into required positions in a 24-well plate; be sure to load samples identically as per **step 2** (*see* **Note 11**).
6. Fill the vacuum workstation as per Fig. 1.
7. Once the 10 min orbit mixing has completed, transfer the plates from **steps 4** and **5** to the vacuum workstation (*see* **Note 12**).
8. Turn on the vacuum workstation and handheld device. Aspirate the bead containing solution until no solution remains and beads are secured to the handheld device (the beads will be visible on the surface of the filters at the end of the each probe).
9. Transfer the handheld device to each trough in the vacuum workstation in numerical order (Fig. 1). The tool should be moved after 5 s in troughs 1 and 2 and 10 s in trough 3. After trough 3, hold the tool vertical and then turn the vacuum valve

Table 2

Dispensation orders for *B. anthracis* SNP assays and Shiga-toxin gene subtyping pyrosequencing.

Target	Dispensation order
B. anthracis	
A/B.Br.001	TATAGAGATAGCGCTGTGATA
A.Br.001	ATCGACTCA GTC GTCTAGATATGT
A.Br.0002	CAGCTACTATACATACTGCTCT
A.Br.003	GCTGAGTAGAGAGTATACATGAC
A.Br.004	ACTGATGTCTATCTCATAGCTCACAT
A.Br.006	TCGCTACGTGCATGTATGATGACTACGCT
A.Br.007	ATGATAGATGTATATCAGTATCGTCTCGTACA
A.Br.008	TCTCTCGCTGTACGTATATG
A.Br.009	GCTGACTGCATCATCGCAGTATATACT
B.Br.001	ATAGAGAGATAGAGTATCGTATCATAT
B.Br.002	CAGCAGTACGTACACGACACAGAGTG
B.Br.003	ATATCAGCTCACGTATGTCGTATGACT
B.Br.004	GATGATAGAGTAGAGTACAGCACT
E. coli	
stx1	GTACAGTGACTGAGACTGAGTAGCTAGTCTGCTGACTATCATGACAGACTCTGTCGTGTA
stx2-bcd	GTAGTATGATGATCATCTCATATACTGACTGACATAGCTAGCAGCATCGCGCA
stx2-aefg	GTAGTCAGTCGTAGTCATCAGTCAGTCAGTATACGATGACTCATGACTGACGTAGCT

Table 3

Pyrosequencing master mix for the immobilization of PCR products.

Reagent	Volume (μL)
Sepharose beads	2
Binding buffer	40
DNase-/RNase-free water	18–33
Biotin labeled PCR product	5–20

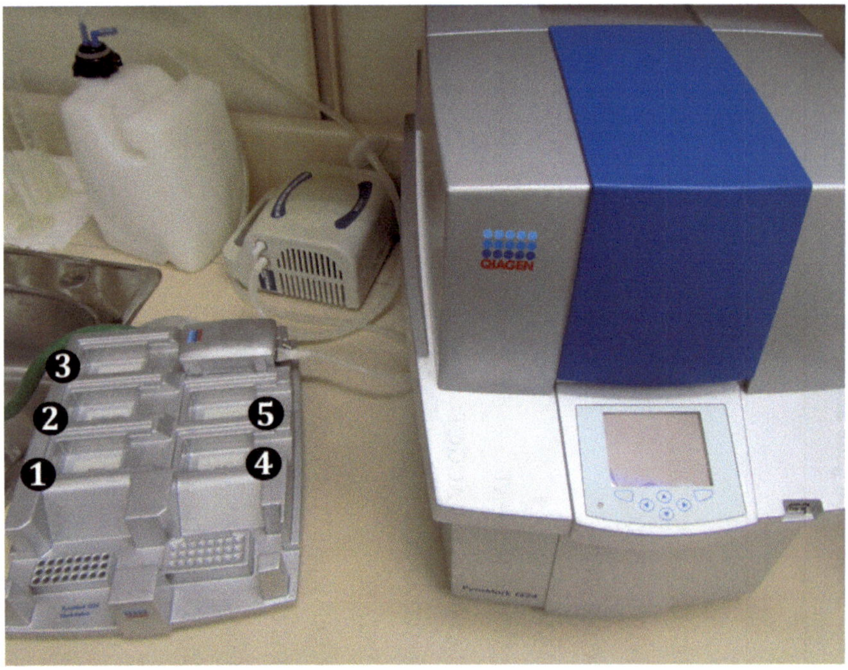

Fig. 1 Pyromark Q24 and vacuum station showing trough positions on vacuum station containing: (*1*) 50 mL of 70 % ethanol, (*2*) 40 mL of denaturation solution, (*3*) 50 mL of wash buffer, (*4, 5*) 50 and 70 mL of high-purity water

to the off position such that the beads can be deposited into the sequencing plate by gentle agitation. After depositing the beads, agitate the tool in trough 4 and turn the vacuum valve open for 30–60 s in trough 5. Lastly, place the vacuum tool in the park position (i.e. the position void of any buffer tray) and turn the vacuum off.

10. Cover the plate containing beads deposited in annealing buffer and sequencing primer with sealing tape and place on a heat block set at 80 °C for 2 min.

11. After 2 min, remove plate from the heat block, incubate at room temperature for 5 min and remove sealing tape.
12. Prepare reagent cartridge by adding the enzyme, substrate, and nucleotides to their respective positions (*see* **Note 13**). Load the reagent cartridge and plate into the PyroMark Q24 system.
13. Turn on the PyroMark Q24 system, insert USB stick containing the run file (generated Subheading 3.3, **step 9**), locate it on the PyroMark Q24 screen and click run (*see* **Note 14**).
14. Following the run, the pyrosequencing data is automatically copied to the USB drive into the run file. A pyrogram of each sample will appear on the information screen of the instrument as pyrosequencing reaction progresses.

3.5 Data Analysis and Interpretation

3.5.1 SNP Analysis of Bacillus anthracis for Resolution of canSNP Phylogeny

The example below describes how pyrosequencing can be used to type *B. anthracis* spores based on their canSNP identities. First, oligonucleotide primers were designed using Geneious to target the 12 previously identified canSNP regions [24]. These primers were then imported into the Qiagen PyroMark Q24 software to generate an appropriate sequencing primer (*see* **Note 15**). The primers were determined to be free of nonspecific interactions and the biotinylated primer was chosen opposite the sequencing primer (*see* Table 4 for the complete primers list). In the example presented (A.Br.002), the SNP exists at the fifth base downstream of the sequencing primer and can occur as either a guanosine or adenosine nucleotide. Fig. 2 illustrates the typical pyrograms observed when sequencing this target (*see* **Note 16**). As can be seen, there is a clear guanosine peak present in the first example (A) while the second example (B) contains a larger adenosine peak representing two nucleotides with an absent guanosine peak. Pyrosequencing this target unambiguously determines the SNP type for this target/isolate. This process is repeated for all SNP loci and the sequencing reads are concatenated into a single read for each organism. Given the high genomic stability of *B. anthracis*, a phylogenetic tree can be constructed based on the concatenated sequences to determine the canSNP profile for each isolate. Known reference sequences (e.g. *B. anthracis* Ames ancestor) can also be included in the comparison. Fig. 3 illustrates a phylogenetic tree generated using Geneious for 28 Canadian *B. anthracis* isolates including 13 from the 2006 Saskatchewan outbreak and 2 reference sequences based on canSNP type. The majority of Canadian isolates contain the same profile and fit in the WNA/A.Br.009 branch. Further, the isolates from the 2006 outbreak (indicated by star symbol) group together as expected. Using this type of analysis it is possible to relate any new outbreak isolates to previous investigations and perform subsequent metadata analysis (e.g. Region isolated, year, season, host organism).

Table 4

PCR primers and pyrosequencing primers for SNP typing of *Bacillus anthracis*

Primer names	Sequence
A/B.Br.001_fwd	AGG CAA TGG ACT GAA TAA AAC G
A/B.Br.001_rvs	/5Biotin/ TGA ACC TTT CGG TAA ATA GTC CC
A.Br.001_fwd	/5Biotin/ TAA GGC AAG CGG AAC CAA AT
A.Br.001_rvs	TCC TGA AAT AAA TTC ACC GTA CGT
A.Br.002_fwd	/5Biotin/ ATT TAT TGG CGG AGT TGC TTC
A.Br.002_rvs	ACC TAA AAT CGA TAA AGC GAC TGC
A.Br.003_fwd	/5Biotin/ AGA AAG TGG TAG AAG CGG TGA AAA
A.Br.003_rvs	CCT GTT CTC AAG TCC CAA AAC ATT
A.Br.004_fwd	CCG ATA CCA GTA AAC GAC GAC AT
A.Br.004_rvs	/5Biotin/ CTG GAA TTG GTG GAG CTA TGG
A.Br.006_fwd	CCG GAA ATT GCT ATT AGA ACG AA
A.Br.006_rvs	/5Biotin/ AGC GTT TTT AAG TTC ATC ATA CCC
A.Br.007_fwd	TGG CGA TTG CGA AAA GTA T
A.Br.007_rvs	/5Biotin/ TTG GTA ACG AGA CGA TAA ACT GA
A.Br.008_fwd	/5Biotin/ ACG TGG GAT GCA AAT AAA CC
A.Br.008_rvs	CAC CGC CAG AAG CTA AGA AA
A.Br.009_fwd	TCG GCC ACT GTT TTT GAA C
A.Br.009_rvs	/5Biotin/ CGG GGT TTC TAC TGT GTA TGT TG

(continued)

Table 4
(continued)

Primer names	Sequence
B.Br.001_fwd	GAA GTT ATT TGC ACG GTC ATA AAA
B.Br.001_rvs	/5Biotin/ AAT TGT TCA AAA GGT TCG GAT ATG
B.Br.002_fwd	AAG AAC AAA ACC GTG TTA GTG ATG
B.Br.002_rvs	/5Biotin/ AGT AGA TTG TTG CAC CTT CTG TGT
B.Br.003_fwd	AGT ATA GCG ATG GTC AAT TCA ATG
B.Br.003_rvs	/5Biotin/ TGC CAT CAA ATA ACT CTT TCT CAA
B.Br.004_fwd	/5Biotin/ AAT TAA TGA TAA AGC GCA AGG TG
B.Br.004_rvs	GCC TTG AGC TTG GTT TAA TAA GA
A/B.Br.001_seq	CAA TCG CTG CAC TCT
A.Br.001_seq	ATA CGG TTT CCC TTT ATC
A.Br.002_seq	ATA AAG CGA CTG CCG
A.Br.003_seq	AAA GCT TGG CAA GCG
A.Br.004_seq	CGA CAT CGC CGT CAT
A.Br.006_seq	TGT TGT TGA TCA TTC CA
A.Br.007_seq	GTG GTA GTA TTC GAG CTG
A.Br.008_seq	ACG TTT TAG ATG GAG ATA AT
A.Br.009_seq	CCA CTG TTT TTG AAC G
B.Br.001_seq	CAT AAA AGA AAT CGG TAC A
B.Br.002_seq	AGA AGT TGC AAA AGG AA
B.Br.003_seq	ATA GAA GCA GAT GAG CTT AC
B.Br.004_seq	GAA GAT AAT GAC AAA CGG

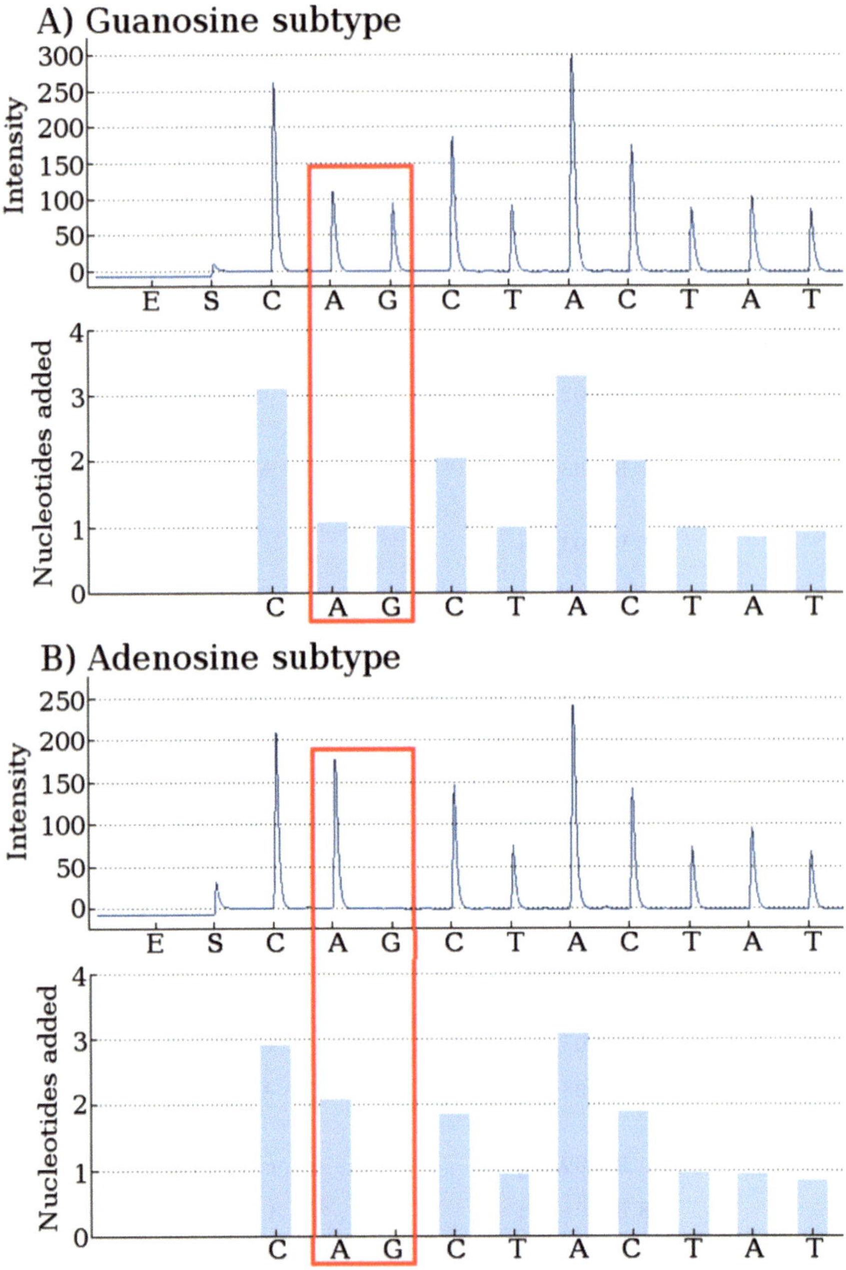

Fig. 2 Pyrosequencing data (target A.Br.002) discriminates between two different SNP subtypes. Subtype (**a**) contains an adenosine nucleotide at base 5 (CCCA**A**CCTAAACCTATAACA), while subtype (**b**) contains a guanosine nucleotide (CCCA**G**CCTAAACCTATAACA)

3.5.2 Stx Detection and Subtyping of Escherichia coli

It is documented that some STEC isolates possess multiple Shiga-toxin genes of different subtypes (e.g. *E. coli* O157:H7 strain Sakai possesses both *stx*-1a and *stx*-2a) [25]. To accomplish amplification of several Shiga-toxin genes simultaneously and to distinguish the

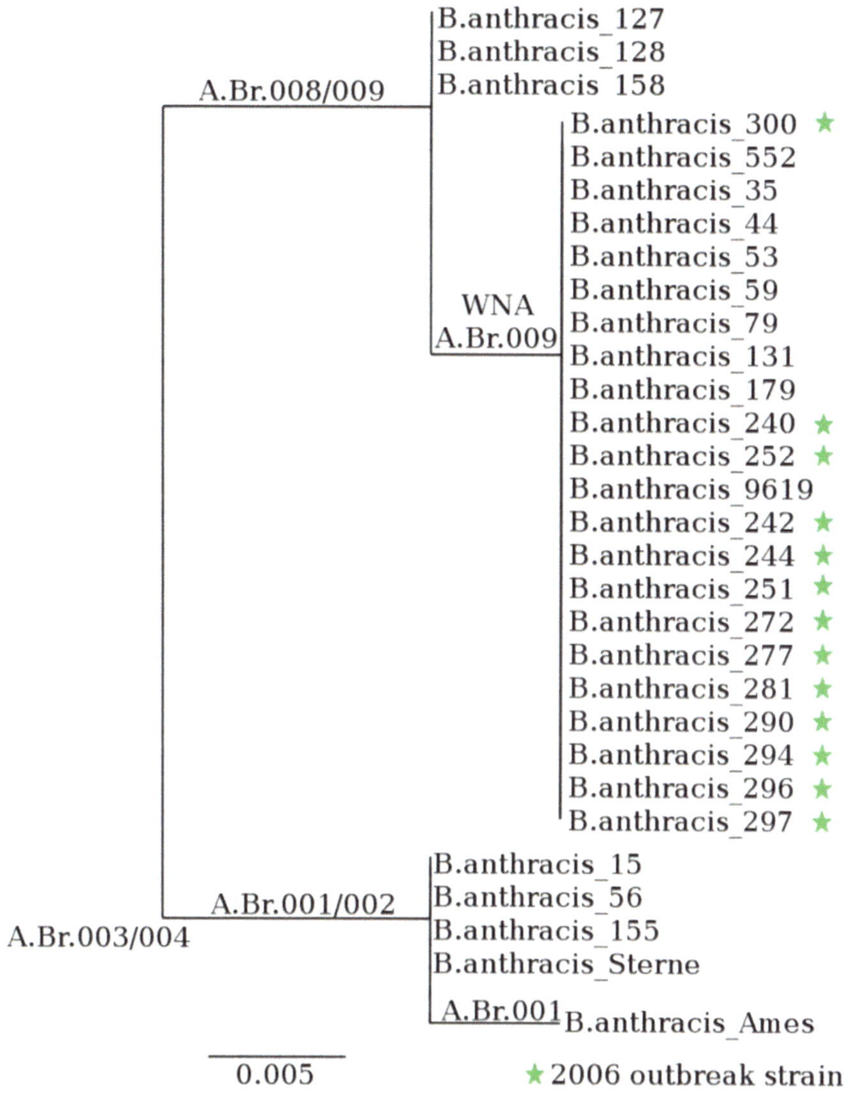

Fig. 3 A phylogenetic tree generated using Geneious R8 illustrates the 28 Canadian *B. anthracis* and 2 reference sequences typed based on canonical SNPs. As expected, the 2006 outbreak strains are all the same type and group together

key subtypes (i.e. *stx*-1a, *stx*-2a, *stx*-2c), each conserved region of the *stx*-1 and *stx*-2 genes were identified by multiple sequence alignments using the Geneious R8 software (Fig. 4 illustrates this for *stx*-2). Subtypes were determined by the characteristic SNPs of each subtype shown in Fig. 4 for *stx*-2 and Fig. 5 for *stx*-1. In order to amplify multiple subtypes from a single isolate and to determine the combination of subtypes, a multiplex PCR assay was designed with one forward primer each for either *stx*-1 and *stx*-2 gene and one reverse primer for *stx*-1 and three reverse primers for *stx*-2

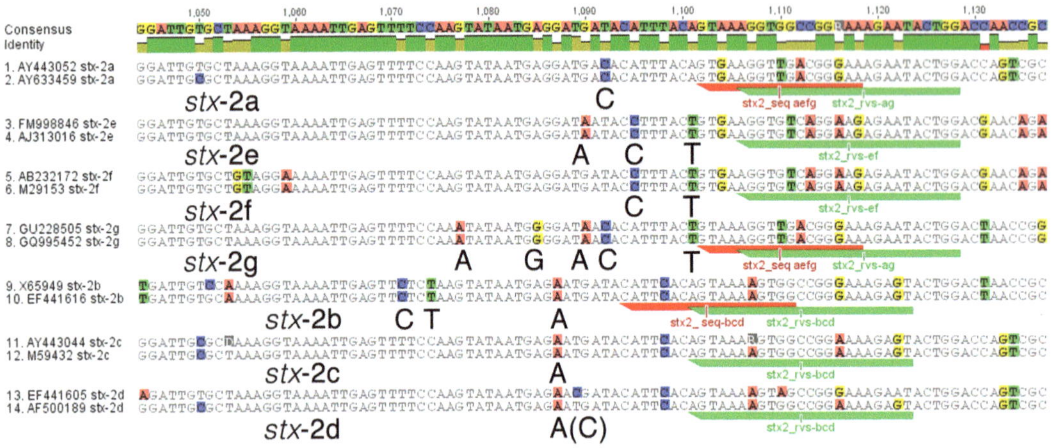

Fig. 4 Multiple alignment image of all Shiga-toxin 2 subtypes (*stx*-2a, -2b, -2c, -2d, -2e, -2f, and -2g) on Geneious R8 software. The bases highlighted in the figure are the signature SNPs of each subtype. Two sequencing primers (stx2-seq-aefg, stx-seq-bcd) bind to the sense side of the gene, thus the dispensation orders are designed to the reverse complement sequence (see dispensation order on Table 2) and sequence reads start with A/T-G-T-AAA for *stx*-2a, -2e, -2f, and -2g while T-A-T-C-G/A for *stx*-2b, -2c, and -2d

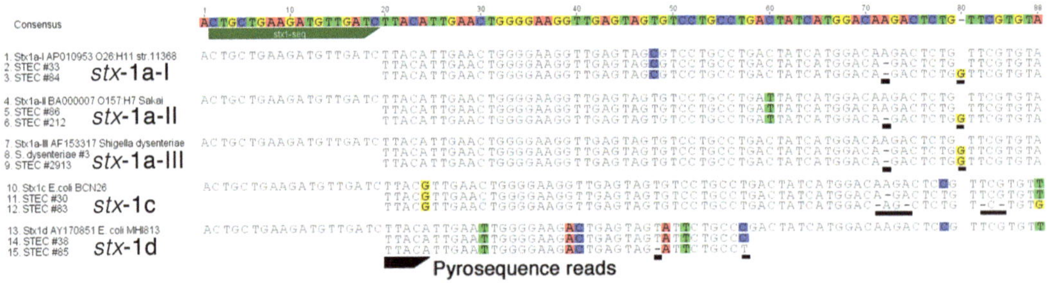

Fig. 5 Multiple alignment image of all Shiga-toxin 1 subtypes (*stx*-1a, -1c, and -1d) with subgroup of *stx*-1a (*stx*-1a Type I, II, and III) and pyrosequencing results underneath the consensus sequences of each subtype (STEC # and *S. dysenteriae* #). Pyrosequence reads clearly distinguish each subtype by SNP typing. Further into the pyrosequencing reads, the number of base-call errors increases (underlined in the figure)

(**stx2_rvs-bcd** for *stx*-2b, -2c, -2d, **stx2_rvs-ag** for *stx*-2a and -2g and **stx2_rvs-ef** for *stx*-2e and 2f) (Table 5) (*see* **Notes 17** and **18**). A sequencing primer and a dispensation for *stx*-1 types were designed as described in the Method section using PyroMark Q24 Assay Design software and PyroMark Q24 Analysis Software on AQ mode. Due to the complexity of the sequences, the sequencing primers and dispensations were designed manually for *stx*-2-bcd and *stx*-2-aefg (Table 2) (*see* **Note 19**).

The Pyrogram results were imported from the PyroMark Q24 Analysis Software with SQA mode, aligned to the sequences

Table 5

Primers for multiplex PCR and pyrosequencing of Shiga-toxin gene subtyping.

Primer name	Sequence	Amplicon size (bp)	Final conc. (μM)
stx1_fwd	ATC TCA GTG GGC GTT CTT ATG	243	0.2
stx1_rvs	/5Biotin/ CAT CTG CCG GAC ACA TAG AAG		0.2
stx2_fwd	/5Biotin/ ATG TCA GAT **WRY** TGG **M**GA CAG G	–	**0.5**
stx2_rvs-**bcd**	AYT CTT TYC CGG CCA CTT TTA CT	269	0.2
stx2_rvs-**ag**	CCA GTA TTC TTT CCC GTC AAC CT	273	0.2
stx2_rvs-**ef**	CCA GTA TTC TCT TCC TGA CAC CT	274	0.2
stx1_seq	CTG CTG AAG ATG TTG ATC	–	0.3
stx2_seq-**bcd**	CCA CTT TTA CTG TGA ATG	–	0.3
stx2_seq-**aefg**	**YTY** CC**K KMM** ACC TTY AC	–	**0.6**

representing each subtype in Geneious R8, and analyzed to identify each subtype based on the SNP pattern. The alignment of the pyrosequencing results to the consensus sequences of each *stx*-1 subtype is shown in Fig. 5.

4 Notes

1. Other master mixes may be used and some, such as the Roche Probes Master kit, have yielded better pyrosequencing results when run in multiplex qPCR and pyrosequencing [3].
2. When using the boiling method for DNA extraction, the DNA degrades quickly and thus must be stored at 4 °C. If storing samples for more than 1 week, store at –20 °C. Long term storage or multiple freeze–thaw cycles are not recommended.
3. Homopolymers are particularly challenging as pyrogram peak interpretation becomes more challenging as the sequencing progresses. This is due to changing signal intensity resulting from dilution of the sample. In some cases, it may be necessary to manually analyze some pyrograms.
4. When preparing a PCR master mix, it is important to do so in an area separate from where DNA template is used to prevent cross-contamination. An ideal scenario is to use a separate room. Both rooms should also be separated from the amplicon-generating area and where the pyrosequencing is completed.

5. Ensure non-biotinylated PCR primers are used for the internal control such that the amplicon does not interfere with downstream pyrosequencing. An internal control ensures DNA was extracted properly, is of sufficient quality, and was actually added to the reaction tube.
6. It is critical that PCR products are not taken back into the PCR setup areas for analysis as this can lead to cross-contamination in future runs.
7. Read quality can be improved if dispensations are optimized by dispensing once for each nucleotide that occurs back to back rather than multiple dispensations of the same nucleotide.
8. SQA assays are often used for sequence confirmation while AQ assays are better suited for SNP detection.
9. The instrument method can be found on the PyroMark cartridge. If the drop-down menu does not display the method then the method file must be obtained from the Qiagen website (https://www.qiagen.com/ca/shop/automated-solutions/pyrosequencing/pyromark-q24/#resources).
10. It is important to use orbital shaker with 3 mm orbit size as vigorous shaking may disrupt the linkage between biotin and streptavidin. Pyrosequencing results should not be affected by extended shaking time up to 15 min shaking time up to 15 min (unpublished data) or potentially longer (untested).
11. This is especially crucial when running multiple different pyrosequencing reactions simultaneously and can be a common source of error. If the plates are not loaded identically, the sequencing primer will likely fail to bind to the PCR amplicon and there will be no signals on the pyrogram.
12. If more than 10 min has elapsed, ensure beads have not settled to the bottom of the plate as this may reduce capture resulting in decreased signal intensity. Put the plate back on the shaker for 3–5 min such that the beads are evenly distributed in solution.
13. Failure to load reagents in the proper order can result in run failure or unexpected sequence results.
14. The first two peaks of the pyrogram are control injections (enzyme/substrate) which will not be included in the output sequence. If there is no peak when the substrate is dispensed then there is a problem with the enzyme and/or substrate or they were not loaded into the correct cartridge positions.
15. The non-biotinylated primer can be used as a sequencing primer, provided the sequence of interest is close enough (<20 bp) to the 3′ end of the primer.
16. Peak calling from the pyrogram is based on 'area under the curve' analysis, not peak height, so the pyrogram scale is not

directly related to the number of bases incorporated in each injection.

17. Primers should be checked for dimers and mispriming using PyroMark Q24 Assay Design software. For multiplex PCRs, it is strongly recommended to use the primer design software to avoid primer dimers. Use non-biotinylated primers first for optimization and validation of the PCR condition before proceeding to pyrosequencing. This will reduce the cost for developing new pyrosequencing assays, since the biotinylated primers need to be ordered HPLC purified to ensure the removal of free biotin.
18. With the limitation in amplicon size (the optimal length of PCR amplicon is from 80 to 300 bp, unpublished data) and the challenge of using multiplexed PCR, primer sequences for *stx*-2 were manually designed and the PyroMark Assay Design software was used only to investigate potential problems caused by hairpin loop of the primers and template, mispriming sites, and duplex formation. NCBI BLAST [21] was also performed to verify the specificity of the primers.
19. In order to acquire longer sequence reads, the best dispensation order should be designed to follow only the consensus sequences with the known SNPs for each *stx*-2 subtypes (Fig. 4). At the same time, it is considerable to insert some extra bases intermittently to prepare for a potential unpublished sequence, especially when targeting the variable region of the gene to be sequenced.

Acknowledgement

The authors thank Dr. Elizabeth Golsteyn-Thomas and Susan Druhan for providing *B. anthracis* isolates. We also would like to acknowledge the technical contribution of Kristen Hahn and Zhen Zhong for the *B. anthracis* pyrosequencing work.

References

1. Ronaghi M, Uhlén M, Nyrén P (1998) A sequencing method based on real-time pyrophosphate. Science 281:363–365
2. Amoako KK, Thomas MC, Kong F et al (2012) Rapid detection and antimicrobial resistance gene profiling of *Yersinia pestis* using pyrosequencing technology. J Microbiol Methods 90:228–234
3. Janzen TW, Thomas MC, Goji N et al (2015) Rapid detection method for *Bacillus anthracis* using a combination of multiplexed real-time PCR and pyrosequencing and its application for food biodefense. J Food Prot 78:355–361
4. Goji N, Mathews A, Huszczynski G et al (2015) A new pyrosequencing assay for rapid detection and genotyping of Shiga toxin, intimin and O157-specific rfbE genes of *Escherichia coli*. J Microbiol Methods 109:167–179
5. Van Ert MN, Easterday WR, Simonson TS et al (2007) Strain-specific single-nucleotide polymorphism assays for the *Bacillus anthracis* Ames strain. J Clin Microbiol 45:47–53

6. Easterday WR, Van Ert MN, Simonson TS et al (2005) Use of single nucleotide polymorphisms in the plcR gene for specific identification of *Bacillus anthracis*. J Clin Microbiol 43:1995–1997
7. Stephens AJ, Huygens F, Inman-Bamber J et al (2006) Methicillin-resistant *Staphylococcus aureus* genotyping using a small set of polymorphisms. J Med Microbiol 55:43–51
8. U'Ren JM, Van Ert MN, Schupp JM et al (2005) Use of a real-time PCR TaqMan assay for rapid identification and differentiation of *Burkholderia pseudomallei* and *Burkholderia mallei*. J Clin Microbiol 43:5771–5774
9. Wahab T, Hjalmarsson S, Wollin R et al (2005) Pyrosequencing *Bacillus anthracis*. Emerg Infect Dis 11:1527–1531
10. Amoako KK, Janzen TW, Shields MJ et al (2013) Rapid detection and identification of *Bacillus anthracis* in food using pyrosequencing technology. Int J Food Microbiol 165:319–325
11. Keim P, Smith KL, Keys C et al (2001) Molecular investigation of the Aum Shinrikyo anthrax release in Kameido, Japan. J Clin Microbiol 39:4566–4567
12. Vogler AJ, Busch JD, Percy-Fine S et al (2002) Molecular analysis of rifampin resistance in *Bacillus anthracis* and *Bacillus cereus*. Antimicrob Agents Chemother 46:511–513
13. Keim P, Van Ert MN, Pearson T et al (2004) Anthrax molecular epidemiology and forensics: using the appropriate marker for different evolutionary scales. Infect Genet Evol 4:205–213
14. Pearson T, Busch JD, Ravel J et al (2004) Phylogenetic discovery bias in *Bacillus anthracis* using single-nucleotide polymorphisms from whole-genome sequencing. Proc Natl Acad Sci U S A 101:13536–13541
15. Moorhead SM, Dykes GA, Cursons RT (2003) An SNP-based PCR assay to differentiate between *Listeria monocytogenes* lineages derived from phylogenetic analysis of the sigB gene. J Microbiol Methods 55:425–432
16. Scheutz F, Teel LD, Beutin L et al (2012) Multicenter evaluation of a sequence-based protocol for subtyping Shiga toxins and standardizing Stx nomenclature. J Clin Microbiol 50:2951–2963
17. Beutin L, Miko A, Krause G et al (2007) Identification of human-pathogenic strains of Shiga toxin-producing *Escherichia coli* from food by a combination of serotyping and molecular typing of Shiga toxin genes. Appl Environ Microbiol 73:4769–4775
18. Persson S, Olsen KE, Ethelberg S et al (2007) Subtyping method for *Escherichia coli* shiga toxin (verocytotoxin) 2 variants and correlations to clinical manifestations. J Clin Microbiol 45:2020–2024
19. Beutin L, Kruger U, Krause G et al (2008) Evaluation of major types of Shiga toxin 2E-producing *Escherichia coli* bacteria present in food, pigs, and the environment as potential pathogens for humans. Appl Environ Microbiol 74:4806–4816
20. Prager R, Fruth A, Busch U et al (2011) Comparative analysis of virulence genes, genetic diversity, and phylogeny of Shiga toxin 2g and heat-stable enterotoxin STIa encoding *Escherichia coli* isolates from humans, animals, and environmental sources. Int J Med Microbiol 301(3):181–191
21. NCBI Resource Coordinators (2014) Database resources of the National Center for Biotechnology Information. Nucleic Acids Res 42:D7–D17
22. Thomas MC, Shields MJ, Hahn KR et al (2013) Evaluation of DNA extraction methods for *Bacillus anthracis* spores isolated from spiked food samples. J Appl Microbiol 115(1):156–162
23. Hahn KR, Janzen TW, Thomas MC et al (2014) Single nucleotide repeat analysis of *B. anthracis* isolates in Canada through comparison of pyrosequencing and Sanger sequencing. Vet Microbiol 169:228–232
24. Van Ert MN, Easterday WR, Huynh LY et al (2007) Global genetic population structure of *Bacillus anthracis*. PLoS One 2:e461
25. Perna NT, Plunkett G III, Burland V et al (2001) Genome sequence of enterohaemorrhagic *Escherichia coli* O157:H7. Nature 409:529–533

Chapter 16

Methods for Genotyping-by-Sequencing

Beth A. Rowan, Danelle K. Seymour, Eunyoung Chae, Derek S. Lundberg, and Detlef Weigel

Abstract

A major goal for biologists is to understand the connection between genes and phenotypic traits, and genetic mapping in experimental populations remains a powerful approach for discovering the causal genes underlying phenotypes. For genetic mapping, the process of genotyping was previously a major rate-limiting step. Modern sequencing technology has greatly improved the resolution and speed of genetic mapping by reducing the time, labor, and cost per genotyping marker. In addition, the ability to perform genotyping-by-sequencing (GBS) has facilitated large-scale population genetic analyses by providing a simpler way to survey segregating genetic variation in natural populations. Here we present two protocols for GBS, using the Illumina platform, that can be applied to a wide range of genotyping projects in different species. The first protocol is for genotyping a subset of marker positions genome-wide using restriction digestion, and the second is for preparing inexpensive paired-end whole-genome libraries. We discuss the suitability of each approach for different genotyping applications and provide notes for adapting these protocols for use with a liquid-handling robot.

Key words Genomic DNA, Reduced-representation, Genetic mapping, RAD-seq, GBS, RESCAN, Sequencing library, Solid phase reverse immobilization

1 Introduction

Genotyping is an essential component of many different avenues of research in biology. For example, whether the investigator is concerned with understanding the genetic basis of phenotypic traits or the processes of molecular evolution, the ease of obtaining genotypes is one factor that affects the number of individuals that can be feasibly included in an experiment.

The cost of high-throughput short-read sequence data has dropped precipitously over the past decade, owing to continued improvements to modern sequencing platforms. The Illumina

Electronic supplementary material: The online version of this chapter (doi:10.1007/978-1-4939-6442-0_16) contains supplementary material, which is available to authorized users.

Stefan J. White and Stuart Cantsilieris (eds.), *Genotyping: Methods and Protocols*, Methods in Molecular Biology, vol. 1492, DOI 10.1007/978-1-4939-6442-0_16, © Springer Science+Business Media New York 2017

platform, in particular, has experienced a dramatic reduction in cost per Gb of sequence over the past few years—from consumable costs of just over \$100/Gb in 2012 to less than \$20/Gb in 2015. One can now obtain genetic information for hundreds, if not thousands, of individuals using genotyping-by-sequencing (GBS), making the investigator able to choose the optimal number of individuals for the experiment. Currently, there are two major strategies for GBS: either performing whole-genome resequencing or reducing the representation of the genome via targeted enrichment or enzymatic digestion before sequencing. With enzymatic digestion, it is more likely that the same protocol can be easily applied across different species, while targeted enrichment strategies are more expensive, less flexible, and would be better suited for researchers working with a single species that has a large genome.

In this chapter, we provide two GBS protocols for the Illumina platform that can be used across different organisms. One is a reduced-representation method using a restriction-site-associated-DNA sequencing strategy (RAD-seq) [1] and the other is a low-cost protocol for whole-genome paired-end resequencing. Each approach has its advantages and limitations, and choosing the correct strategy for any project is of considerable importance and is addressed in Table 1.

Table 1
Comparing strategies for genotyping-by-sequencing

	Reduced-representation (KpnI)	Whole genome
Practical considerations		
Library preparation cost (per sample)[a]	<\$1	\$3–5
Lab time (per multiplex)[b]	<16 h (384-plex)	12–16 h (96-plex)
Suitability for large genomes	More suitable	Less suitable
Number of individuals	Very large	Medium–large
Sequencing cost (per sample)[c]	\$3–4	\$12–15
Project goals		
Quantitative trait locus mapping	Yes	Yes
Mapping interval size[d]	Larger	Smaller
Causal polymorphism identification	No	Yes
Population genetics	Less suitable	More suitable

[a]Costs include fragmentation and reagents for library preparation, but do not include costs for DNA quantification, library validation, or plastic consumables and are based on list prices for the US market

[b]The time required starting from normalized DNA samples to final library validation

[c]Assumes that 384 samples (reduced-representation) or 96 samples (whole-genome resequencing) are pooled in a single HiSeq 2000 lane. Multiple whole genome 96-plexes can be pooled in a single lane of either the HiSeq 2000 or 3000 depending on desired coverage per sample, reducing the sequencing cost to as low as \$2–3 per sample

[d]For the same number of individuals (recombination events)

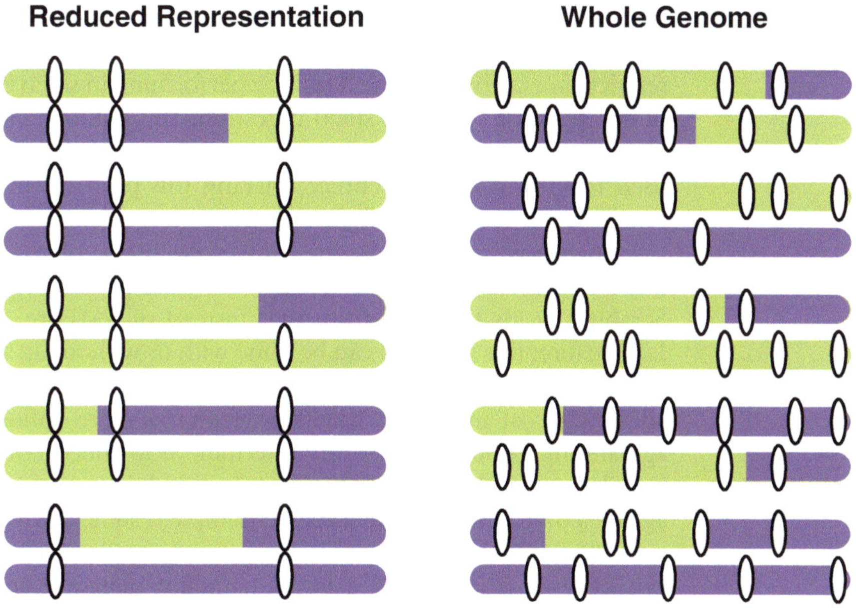

Fig. 1 Comparison of strategies for genotyping-by-sequencing. A schematic representation of a single chromosome pair for five recombinant individuals is shown. The parental genotypes are shown in *purple* and *green*. *White ovals* indicate the sampled markers. For the same total coverage per individual, the reduced representation approach (like the KpnI-based protocol we describe in this chapter) samples fewer markers with nearly complete coverage across indviduals (*left*) compared with the whole genome approach (*right*)

Assuming the same total coverage for 384 individuals, the RAD-seq protocol will sample fewer markers at a higher average coverage, while the whole genome resequencing protocol will have a higher marker density, but lower average coverage per marker (Fig. 1).

The reduced-representation protocol that we present here is based on [1] and [2], and aims to interrogate relatively few markers at high coverage while limiting the amount of missing genotype information per individual, a recurrent problem in other reduced-representation approaches that complicates downstream analyses. We increase the reproducibility of fragment enrichment across libraries by coupling enzymatic digestion to random shearing, removing the need for targeted size selection of fragments. Our chosen restriction enzyme, KpnI, samples about 1 %, or 1.5 Mb, of the *Arabidopsis thaliana* genome; other restriction enzymes might be better suited for other species. Sequencing multiplexed libraries of 384 individuals produces approximately 30× coverage of each sequenced base when pooled in a single lane of the Illumina HiSeq 2000 platform (about 20 Gb total sequence—costing about $65/Gb). Data from previous resequencing experiments [3] predicts that approximately 2000 sites, residing on the sequenced ends of the KpnI fragments, will segregate between any two *A. thaliana* strains, making this protocol ideal for the molecular dissection of phenotypic differences in experimental populations of this species. If more genetic markers are needed, this protocol can be easily adapted to work with any restriction enzyme that produces sticky ends. For species with reference genome sequences, the optimal restriction enzyme can be chosen by performing in silico digestions of the genome. After in silico digestion, the number of genomic fragments generated and the distribution of their lengths can be used to inform enzyme choice, making this protocol transferable to nearly any species of interest.

The whole-genome resequencing protocol we present is adapted from [4]. Here, we further reduce the cost per library by using Serapure beads (adapted from [5]) instead of AMPure XP beads. Like before, this protocol can be done with or without an additional normalization before pooling the libraries. Normalization, through additional cost and effort, will likely result in a more uniform coverage among samples. Users can determine how much coverage per sample is needed for their experiment; we find that an average coverage of 0.5× to 1× of the genome per sample is sufficient for accurate genotype calling. The protocol can be performed entirely by hand, but benefits from use of a liquid-handling pipetting robot. The complete methods for both protocols detailed here are presented as they are performed manually, with suggestions for how to adapt them for a liquid-handling pipetting robot.

2 Materials

2.1 KpnI Reduced-Representation Library Preparation

2.1.1 Preparing Multiple 384-Plex Libraries

1. 50 μM stocks of 384 barcoded oligos (eight 96-well plates total) (Supplementary Table 1). The template adapter must be 5′ phosphorylated.
2. 100 μM stocks of HPLC purified universal adapter oligos (Supplementary Table 1). Universal Adapter 1 must be 5′ phosphorylated.
3. 100 μM stocks of HPLC purified universal PCR oligos (Supplementary Table 1).
4. Twelve Axygen 96-well microplates.
5. Four 1.5-mL Eppendorf microcentrifuge tubes.
6. Sixteen ABgene 96-well plate seals (Thermo Fisher Scientific).
7. 1 mL Sera-Mag SpeedBeads (Thermo Fisher Scientific).
8. TE solution: 10 mM Tris–HCl, 1 mM EDTA, pH 8.
9. PEG-8000.
10. NaCl.
11. 1 M Tris–HCl, pH 8.
12. 0.5 M EDTA, pH 8.
13. Agarose.
14. GeneRuler DNA ladder mix (Thermo Fisher Scientific).
15. 6× gel loading dye aliquots (Thermo Fisher Scientific).

2.1.2 384-Plex Libraries

1. Eight Axygen 96-well microplates.
2. Eight ABgene 96-well plate seals.
3. Two 2.0 mL Eppendorf microcentrifuge tubes.
4. Five 8-strip microtubes or five 1.5-mL Eppendorf tubes.
5. Four black 96-well plates with clear, flat bottoms, or the type of plate suited for the fluorescence plate reader (optional).
6. Quant-It PicoGreen Kit (optional).
7. 384 μL (+10 % volume) of FastDigest KpnI Restriction enzyme (Thermo Fisher Scientific).
8. 384 μL (+ 10 % volume) of T4 DNA Ligase.
9. One reaction from NEBNext DNA Sample Prep Master Mix Set for Illumina.
10. One Covaris snap-cap microTUBE.
11. 45 mL of 10 mM Tris-HCl, pH 8.5 or Qiagen EB Buffer.
12. 350 mL of freshly prepared 80 % molecular grade ethanol.
13. Six 50 mL disposable pipetting reservoirs.
14. 60 mL of Serapure solid phase reversible immobilization (SPRI) beads, prepared as in Subheadings 3.1.4 and 3.1.5.
15. Qubit HS or BR assay kit.
16. Agilent DNA 1000 Kit.

2.1.3 Required Equipment

1. 96-well thermal cycler.
2. Multi-channel flexible volume pipettes (each with a 10, 50, and 200 μL maximum volume).
3. Ambion Magnetic Stand-96 (Thermo Fisher Scientific).
4. NEB 6-tube Magnetic Separation Rack.
5. Covaris S220 Focused-ultrasonicator.
6. Agilent 2100 Bioanalyzer.
7. Qubit Fluorometer.
8. Fluorescence plate reader (optional).

2.2 Low Cost Whole Genome Sequencing Library Preparation

2.2.1 Preparing Multiple 96-Plex Libraries

1. Stocks of HPLC-purified 96 P1 adapter oligos with index sequences with 5′ phosphate and 3′ phosphorothioate modifications (Supplementary Table 2), diluted to 100 μM in pure water (MilliQ or similar).
2. Stocks of HPLC-purified Universal P2 adapter oligos (Supplementary Table 2), diluted to 100 μM in pure water (MilliQ or similar).
3. Stocks of HPLC or salt-free purified PCR oligos with a 3′ phosphorothioate modification (Supplementary Table 3), diluted to 100 μM in pure water (MilliQ or similar).

4. Axygen 96-well microplates.
5. 1.5-mL Eppendorf tubes.
6. ABgene 96-well plate seals.
7. 10× Annealing Buffer (AB): 500 mM NaCl, 100 mM Tris, pH 7.5–8.0 (or Buffer 2 from New England Biolabs).
8. Sera-Mag SpeedBeads (Thermo Fisher Scientific).
9. TE solution: 10 mM Tris–HCl, 1 mM EDTA, pH 8.
10. PEG-8000.
11. NaCl.
12. 1 M Tris–HCl, pH 8.
13. 0.5 M EDTA, pH 8.
14. Agarose.
15. Qubit HS assay kit.
16. Silicone cap mats.
17. High Sensitivity DNA chips and reagent kits for the Agilent Bioanalyzer.
18. GeneRuler DNA ladder mix.
19. 6× gel loading dye aliquots.
20. 15- and 50-mL conical Falcon tubes.
21. Quant-It PicoGreen Kit (optional).
22. Floating tube and plate racks.
23. 5 L Beakers.

2.2.2 96-Plex Libraries

1. Prepared adapter plate with each well containing 5 μM Indexed P1 and 5 μM P2 adapter (*see* Subheading 3.2.1 and Supplementary Table 2).
2. 48 μL (48 U + 10% extra for mastermix) of dsDNA Shearase.
3. 480 μL (+10% extra for mastermix) of 5× dsDNA Shearase buffer: 50 mM Tris–HCl, pH 7.5, 125 mM $MgCl_2$, 5 mM DTT.
4. Elution buffer: 10 mM Tris–HCl, pH 8.5.
5. Freshly prepared 80% molecular grade ethanol.
6. Five or three 96-well Axygen plates (if normalization step is skipped).
7. 5–10 mL Sera Pure beads, washed and prepared according to Subheadings 3.1.4 and 3.1.5.
8. 240 μL (+10% extra for mastermix) of NEB Buffer 2.
9. 48 μL (+10% extra for mastermix) of 10 mM dATP.
10. 48 μL (240 U +10% extra for mastermix) of Klenow exo-.
11. 288 μL (+10% extra for mastermix) of 10 mM ATP.

12. 48 μL (+10% extra for mastermix) of Quick Ligase (NEB).
13. Phusion Taq polymerase, amount varies.
14. 10 mM dNTPs, amount varies.
15. Black 96-well plates with clear, flat bottoms (optional).
16. Axymat silicone sealing mats (Axygen).

2.2.3 Required Equipment

1. 96-well thermal cycler.
2. Multi-channel flexible volume pipettes (each with a 10, 50, and 200 μL maximum volume).
3. Magnetic 96-well plate stand.
4. 6-tube Magnetic Separation Rack NEB.
5. Covaris S220 Focused-ultrasonicator.
6. Agilent 2100 Bioanalyzer.
7. Qubit Fluorometer.
8. Fluorescence plate reader (optional).

3 Methods

3.1 384-Plex Reduced-Representation KpnI Library Preparation

3.1.1 DNA Quantification and Normalization

Quantify DNA on the Qubit with the Qubit HS or BR assay kit, depending on the starting DNA concentration. A plate fluorometer can also be used with the Quant-It PicoGreen Kit or another similar intercalating dye specific for double stranded DNA to quantify the input DNA. To verify DNA quality, perform a test restriction enzyme digest on a subset of samples to ensure that the DNA can be cleaved. Typically, the test is performed with a frequent cutter, such as EcoRI, as KpnI cleaves infrequently and it can be difficult to determine digestion completion. Normalize DNA in a 96-well microplate to ensure that starting material (200 ng in 30 μL) is equal for each sample; otherwise the sequenced 384-plex will have uneven representation among individual samples.

3.1.2 KpnI Barcoded Adapter Preparation

1. Dilute each of the KpnI oligos (*see* Table 2) to 2.5 μM concentration by adding 10 μL of the 50 μM adapter stock to 190 μL of sterile deionized water. There are eight oligo plates total, two for each barcoded adapter sequence. When genotyping fewer than 384 samples *see* **Note 1**. This protocol is for single-end sequencing. To adapt them for paired-end sequencing *see* **Note 2**.
2. Mix template strand (tempad) and sequence strand (seqad) and dilute to 0.05 μM concentration. Add 4 μL of 2.5 μM template strand oligo and 4 μL of 2.5 μM sequence strand oligo to 192 μL of sterile deionized water.
3. Store all oligos at −20 °C and avoid freeze-thaw cycles of 50 and 2.5 μM stocks.

3.1.3 Universal Oligo Preparation

1. For the universal adapters, dilute the stock oligos to 10 μM concentration and mix by adding 10 μL each of the 100 μM adapter stock of Universal Adapter 1 and Universal Adapter 2 to 80 μL of sterile deionized water.
2. Dilute each stock PCR oligo to 10 μM concentration by adding 10 μL of the 100 μM Universal PCR oligo stock to 90 μL of sterile deionized water.

3.1.4 Prepare 100 mL SeraPure SPRI Beads

1. Mix Sera-mag SpeedBeads and transfer 1 mL of beads to a 1.5-mL microcentrifuge tube. Mix beads extremely well before transferring. Vortex the bottle and look carefully to be sure that no beads remain stuck to the bottom and that all beads are fully resuspended.
2. Place SpeedBeads on a magnetic separation rack for 1.5-mL microcentrifuge tubes until beads are drawn to magnet.
3. Remove and discard supernatant.
4. Add 1 mL TE to beads, remove from magnet, mix, and return to magnet.
5. Remove and discard supernatant.
6. Repeat **steps 4** and **5**.
7. Add 1 mL TE to beads and remove from magnet. Fully resuspend and set microtube in rack.
8. Add 18 g PEG-8000 to a sterile bottle capable of storing at least 100 mL.
9. Add nuclease-free water up to the 70-mL mark.
10. Mix by swirling the bottle by hand until well dissolved (10 min).
11. Add 14.6 g NaCl to the bottle in small portions, mixing well each time by swirling.
12. After all of the NaCl has been added, add 1 mL 1 M Tris–HCl and 200 μL 0.5 M EDTA.
13. Mix well and bring volume to 100 mL with nuclease-free water. Final PEG/NaCl solution: 2.5 M NaCl, 20 mM PEG, 10 mM Tris–HCl, 1 mM EDTA.
14. Let the solution rest for a few minutes until bubbles have cleared.
15. Filter using a 0.45- or 0.22-μm bottle top filter if a sterile solution is desired.
16. Add 1 mL of the washed beads to the solution.
17. Divide into two 50-mL Falcon tubes and store at 4 °C.
18. Test the performance of the beads monthly (*see* Subheading 3.1.5) and prepare a new batch of beads if the size distributions are not as expected.

3.1.5 Test SPRI Beads by Performing Bead Clean-Up in Microcentrifuge Tubes

1. Prepare three 1.5-mL microcentrifuge tubes containing 6 μL GeneRuler DNA ladder mix (or similar DNA ladder) mixed with 14 μL water.
2. Add the prepared SPRI beads at 1.8:1, 1:1, and 0.6:1 ratios (36, 20, and 12 μL; 1 tube each).
3. Mix well and incubate at room temperature for 5 min to allow the DNA fragments to bind to the DNA.
4. Transfer tubes to a magnetic separation rack for 1.5-mL microcentrifuge tubes and incubate for 5 min.
5. Remove supernatant while tubes are sitting on the magnetic rack. Do not disrupt the bead pellet.
6. Wash 2× with 700 μL (or the volume necessary to completely cover the beads) of freshly-prepared 80% ethanol. For each wash, add the ethanol, wait 30 s, then remove with a pipet. Do not disrupt the bead pellet.
7. After removing the second ethanol wash, allow beads to dry at room temperature until all of the ethanol has evaporated (5–20 min, *see* **Note 3**).
8. Remove tubes from magnet and resuspend beads in 20 μL 10 mM Tris, pH 8.5. Make sure that the beads are thoroughly mixed with the Tris elution buffer.
9. Incubate tubes at room temperature for 5 min to allow the DNA to come into solution.
10. Transfer tubes to magnet and incubate for 5 min or until all of the beads have been pulled to the magnet and the Tris elution buffer appears clear.
11. Transfer 20 μL of the eluted DNA to a tube.
12. Add 4 μL 6× gel loading buffer to each sample.
13. Load each sample into one well of a 1% agarose gel. Include a lane with a sample prepared with 6 μL of the ladder with 4 μL gel loading buffer and 14 μL 10 mM Tris, pH 8.5.
14. Run gel for 20 min at 8 V/cm.
15. Evaluate the size distribution of fragments (*see* Fig. 2) and the intensity of the bands compared with the ladder sample that did not go through the bead clean-up. Verify that the ladder sample from the SPRI cleanup using the 0.6 ratio does not exhibit bright bands below 500 bp. If this happens, then the ratios for size selection in Subheading 3.2.10 will not yield fragments of the correct size range. If size distributions are not as expected, prepare the SeraPure SPRI beads again (Subheading 3.1.4).

3.1.6 Restriction Enzyme Digestion of DNA with FastDigest KpnI

1. Perform restriction enzyme digest of 200 ng of DNA in a 30 μL volume using 1 μL of FastDigest KpnI enzyme and 3 μL

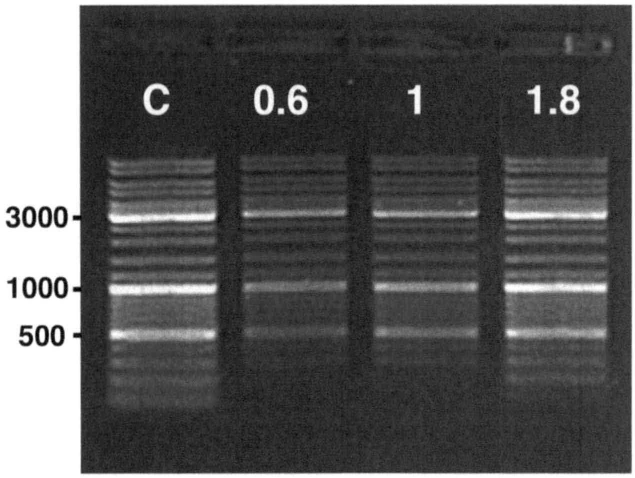

Fig. 2 Gel electrophoresis showing results of size selection and clean-up of DNA fragments using Sera Pure beads. SeraPure beads were used to clean up 0.5 μg of Thermo Fisher Gene Ruler DNA ladder (*see* Subheadings 3.1.4 and 3.1.5) by adding a ratio of either 0.6, 1, or 1.8 volumes Serapure beads to 1 volume of DNA ladder. The cleaned DNA ladder samples were analyzed along with 0.5 μg of a ladder sample that had not been cleaned up (labeled with C) on a 1 % agarose gel at 8 V/cm for 20 min. Numbers on the *left* indicate the size of the DNA fragments in bp.

of the supplied 10× reaction buffer in each well of a 96-well microplate. Add water up to 30 μL volume.

2. Cover with a plate seal and incubate the reactions in a thermal cycler at 37 °C for 30 min.

3.1.7 SPRI Bead Clean-Up in 96-Well Format

1. Add 54 μL (a 1.8:1 ratio) of prepared SPRI beads (*see* Subheading 3.1.4) to each sample. Mix thoroughly by pipetting.
2. Incubate samples at room temperature for 10 min.
3. Place sample plate on 96-well plate magnet for 5 min or until solution is cleared of beads.
4. Remove 75 μL of the supernatant while the sample remains on the magnet. Do not disrupt the beads.
5. Add 200 μL of freshly prepared 80% ethanol to each sample while still on the magnet. Wait for at least 30 s.
6. Remove all of the supernatant from each sample. Do not disrupt the beads.
7. Add 200 μL of freshly prepared 80% ethanol to each sample while still on the magnet. Wait for at least 30 s.

8. Remove all of the supernatant from each sample. Do not disrupt the beads (*see* **Note 3**).
9. Dry the samples at room temperature for up to 20 min or until all ethanol has evaporated. Once the bead pellet is dry, remove the plate from the magnet.
10. Elute sample by adding 12 μL of 10 mM Tris, pH 8.5 to each sample.
11. Mix elution buffer thoroughly with beads (*see* **Note 4**).
12. Leave samples at room temperature for 2 min.
13. Return the plate to the magnet for 5 min or until solution is cleared of beads.
14. Pipet 10 μL of the eluate into a new 96-well microplate.

3.1.8 KpnI Adapter Ligation Using T4 DNA Ligase

1. Perform ligation of digested DNA (after SPRI clean-up) and 1 μL of 0.05 μM KpnI adapters in a 30 μL volume using 3 μL (5 % w/v) of 50 % PEG 4000, 3 μL of the supplied 10× reaction buffer, 12 μL of water, and 1 μL (5U) of T4 Ligase in the 96-well microplate containing digested DNA.
2. Cover with plate seal and incubate the reactions at room temperature for 30 min followed by a 65 °C incubation in a thermal cycler for 20 min to inactivate the enzyme.
3. Place the reactions on ice for 15 min.
4. Add 70 μL of 10 mM Tris to the 30-μL ligation reaction. Mix thoroughly by pipetting.
5. Perform SPRI bead clean-up as in Subheading 3.1.7 using 100 μL of beads (1:1) ratio, 200 μL for each ethanol wash, an elution volume of 17 μL of 10 mM Tris, and transferring 15 μL of eluate into a new 96-well microplate.

3.1.9 Multiplex and Concentrate Adapter Ligated Samples

1. Combine up to 192 barcoded samples (*see* **Note 5**) by pipetting 5 μL of each sample into a 2-mL microcentrifuge tube. Mix thoroughly by pipetting.
2. Perform SPRI bead clean-up in microcentrifuge tube format by following **steps 2** through **11** in Subheading 3.1.5, using 960 μL of beads (1:1) ratio, 1000 μL for each ethanol wash, an elution volume of 30 μL of 10 mM Tris, and transferring 28 μL of eluate into a new 1.5-mL microfuge tube.
3. Combine two 192-plexes by pipetting 28 μL of the eluate from each sample into a new 1.5-mL microcentrifuge tube. The final volume should be 56 μL (*see* **Note 6**).

3.1.10 Fragment the 384-Plex Using Covaris

1. Pipet 55 μL of 384-plex into a new Covaris microTube.
2. Place microTube in Covaris and shear DNA to a mean fragment size of 500 base pairs (Duty cycle: 10%; Intensity: 5; Cycles per burst: 200; Time: 40 s).

3. Transfer sheared sample to new 1.5-mL microcentrifuge tube.
4. Add 45 μL of 10 mM Tris to the 55 μL of sheared DNA. Mix thoroughly by pipetting.
5. Perform SPRI bead clean-up in microcentrifuge tube format by following **steps 2** through **11** in Subheading 3.1.5, using 100 μL of beads (1:1) ratio, 200 μL for each ethanol wash, an elution volume of 22 μL of 10 mM Tris, and transferring 20 μL of eluate into a new 8-well strip tube (*see* **Note 7**).

3.1.11 End Repair of 384-Plex Using NEBNext DNA Sample Prep MMS1

1. Perform end repair of sheared DNA in a 100-μL volume. Combine 20 μL of the sheared DNA with 10 μL of the supplied 10× reaction buffer, 65 μL of water, and 5 μL of the supplied End Repair Enzyme Mix.
2. Incubate the sample in a thermal cycler for 30 min at 20 °C.
3. Perform SPRI bead clean-up as in Subheading 3.1.7, using 100 μL of beads (1:1) ratio, 200 μL for each ethanol wash, an elution volume of 32 μL of 10 mM Tris, and transferring 30 μL of eluate into a new 8-well strip tube.

3.1.12 dA-Tailing of 384-Plex Using NEBNext DNA Sample Prep MMS1

1. Perform dA-tailing of end repaired DNA in a 50 μL volume. Combine 30 μL of the end repaired DNA with 5 μL of the supplied 10× reaction buffer, 12 μL of water, and 3 μL of the supplied Klenow Fragment (3′–5′ exo).
2. Incubate the sample in a thermal cycler for 30 min at 37 °C.
3. Add 50 μL of 10 mM Tris to the 50-μL reaction. Mix thoroughly by pipetting.
4. Perform SPRI bead clean-up as in Subheading 3.1.7, using 100 μL of beads (1:1) ratio, 200 μL for each ethanol wash, an elution volume of 32 μL of 10 mM Tris, and transferring 30 μL of eluate into a new 8-well strip tube.

3.1.13 Universal Adapter Ligation Using NEBNext DNA Sample Prep MMS1

1. Perform ligation of dA-tailed DNA in a 50-μL volume. Combine 30 μL of the dA-tailed DNA with 1 μL of universal adapter (10 μM), 10 μL of the supplied 5× reaction buffer, 4 μL of water, and 5 μL of the supplied Quick T4 Ligase.
2. Incubate the sample in a thermal cycler for 15 min at 20 °C.
3. Add 50 μL of 10 mM Tris to the 50 μL reaction. Mix thoroughly by pipetting.
4. Perform SPRI bead clean-up as in Subheading 3.1.7, using 100 μL of beads (1:1) ratio, 200 μL for each ethanol wash, an elution volume of 27 μL of 10 mM Tris, and transferring 25 μL of eluate into a new 8-well strip tube.

3.1.14 PCR Enrichment Using NEBNext DNA Sample Prep MMS1

1. Perform PCR of ligated DNA in a 30-μL volume. Combine 5 μL of the ligated DNA with 1 μL of the forward and reverse PCR primers (10 μM), 8 μL of water, and 15 μL of the supplied 2× Phusion HF Master Mix.
2. Incubate the sample in a thermal cycler using the following cycling protocol: 98 °C for 30 s; 12–14 cycles of 98 °C for 10 s, 65 °C for 30 s, 72 °C for 30 s; 72 °C for 5 min; hold at 10 °C (*see* **Note 8**).
3. Add 70 μL of 10 mM Tris to the 30 μL reaction. Mix thoroughly by pipetting.
4. Perform SPRI bead clean-up as in Subheading 3.1.7 using 100 μL of beads (1:1) ratio, 200 μL for each ethanol wash, an elution volume of 27 μL of 10 mM Tris, and transferring 25 μL of eluate into a new 8-well strip tube.

3.1.15 Library Validation

1. Measure final library with the Qubit BR assay and validate concentration and size distribution on the Agilent Bioanalyzer (use a DNA 1000 chip). *See* Fig. 3a for a Bioanalyzer trace of a successful library.
2. For sequencing, dilute the final library to 10 nM and validate concentrations with the Qubit BR or HS assay. Use the concentration (ng/μL) and mean fragment size obtained from the Agilent Bioanalyzer trace to calculate nanomolarity (nM).

3.2 Low Cost 96-Plex Whole-Genome-Sequencing Library Preparation

3.2.1 Preparation of Indexed and Universal Adapters

1. Prepare 100 μM stocks for each single stranded oligonucleotide listed in Supplementary Table 2 in 10 mM Tris, pH 8.
2. Combine complementary sense and antisense indexed adapter oligos at 10 μM in 200 μL of 1× Annealing Buffer (AB) in a 96-well microplate.
3. Combine sense and antisense oligos for the P2 Universal Adapter at 10 μM in 5 mL of 1× AB in a 5-mL microfuge tube or a 15-mL conical tube.
4. Divide the 5 mL of 10 μM sense and antisense oligos for the P2 Universal Adapter into ten 1.5-mL microcentrifuge tubes so that each contains 0.5 mL of the oligo mix.
5. Place the P2 Adapter tubes and the 96-well plate containing the indexed adapter oligos in floating racks in a beaker of water that has been heated to the point of boiling and removed from the heat source.
6. Allow samples to cool slowly to room temperature to anneal (4 h or overnight).
7. Mix each of the annealed 96 indexed adapters with the annealed P2 Universal Adapter at a concentration of 5 μM for a total

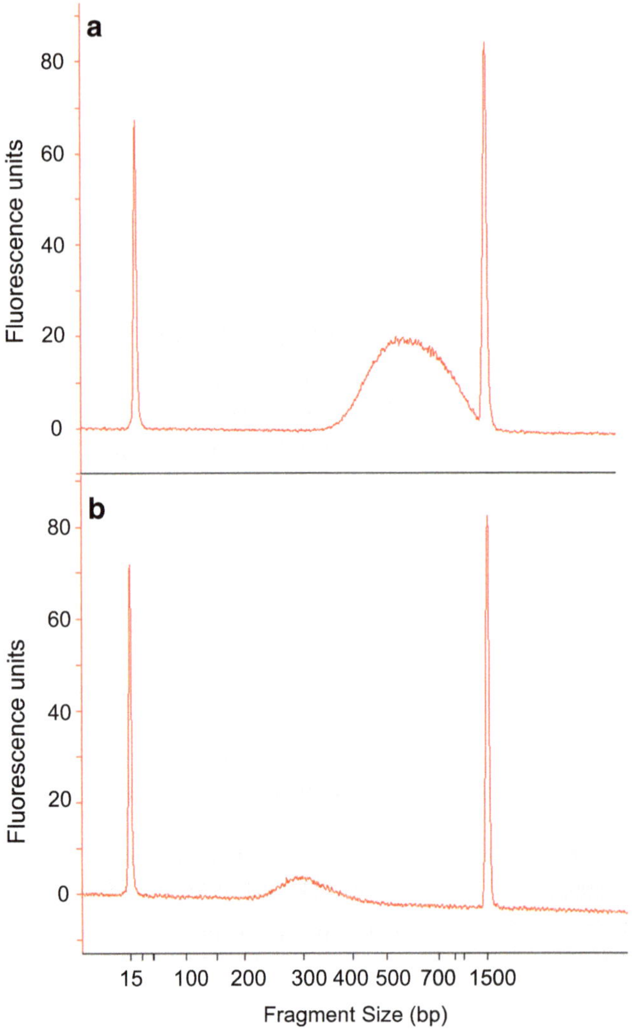

Fig. 3 Bioanalyzer validation of final pooled libraries. (**a**) KpnI reduced representation sequencing 384-plex library. (**b**) Whole genome sequencing 96-plex library. The size distribution shown in (**a**) is obtained without size selection. If smaller fragments are needed, an optional size selection step can be performed

volume of 100 μL in each well of a 96-well plate. If you use the well positions listed in Table 3, this will facilitate pooling when fewer than 96 samples are analyzed (*see* **Note 9**). This should provide enough adapters to make approximately 100 96-plex libraries.

3.2.2 PCR Primer and SPRI Bead Preparation

1. Prepare 100 μM stocks for the forward and reverse primers (Supplementary Table 3) in pure, sterile water (MilliQ or similar).

2. Dilute and prepare aliquots at 10 μM in pure, sterile water (MilliQ or similar).
3. Prepare SPRI beads as in Subheadings 3.1.4 and test them as in 3.1.5.

3.2.3 Quantification and Normalization of DNA Samples

1. Start with very pure DNA samples that are of high molecular weight (>10 kb). Standard laboratory protocols such using cetyltrimethylammonium bromide (CTAB) followed by chloroform extraction and ethanol precipitation are usually sufficient.
2. Quantify DNA on the Qubit with the Qubit HS assay kit or on a plate fluorometer using the Quant-It PicoGreen Kit or another kit with an intercalating dye specific for double stranded DNA. Normalize DNA to 100 ng in 15.5 μL of 10 mM Tris in a 96-well microplate. If necessary, inputs as low as 50 ng can be used.

3.2.4 Fragmentation of DNA Using dsDNA Shearase

1. Make a reaction mix for all 96 samples (prepare an additional 10% extra mix to ensure there is enough) containing the following components per sample:

5× Shearase buffer	5 μL
Water	4 μL
Shearase enzyme	0.5 μL (0.5 U)

2. Pipet 9.5 μL of the reaction mix into each well of the normalized DNA plate for a total reaction of 25 μL.
3. Cover the plate with a silicone cap mat and incubate reactions in a thermal cycler at 42 °C for 30–35 min, followed by 5 min at 65 °C to inactivate the Shearase. *See* **Note 10** if using species other than *Arabidopsis thaliana*.
4. Perform SPRI bead clean-up as in Subheading 3.1.7, using 45 μL of beads (1.8:1) ratio, 200 μL for each ethanol wash, an elution volume of 18 μL of 10 mM Tris, and transferring 17 μL of eluate into a new 96-well microplate.

3.2.5 dA-Tailing of Fragments Using NEB Klenow exo-

1. Make a reaction mix for 96 samples (prepare an additional 10% extra mix to ensure there is enough) containing the following components per sample:

Water	4.5 μL
NEB Buffer 2	2.5 μL
10 mM dATP	0.5 μL
Klenow exo -	0.5 μL

2. Pipet 8 μL of the reaction mix into 17 μL fragmented DNA from Subheading 3.2.4.
3. Cover the plate and incubate at 37 °C for 30 min, then heat inactivate at 75 °C for 20 min in a thermal cycler.
4. Perform SPRI bead clean-up as in Subheading 3.1.7, using 45 μL of beads (1.8:1) ratio, 200 μL for each ethanol wash, an elution volume of 12 μL of 10 mM Tris, and transferring 11 μL of eluate into a new 96-well microplate (*see* **Note 6**).

3.2.6 Ligation of Barcoded Adapters Using NEB T4 DNA Ligase

1. Pipet 1 μL of the 5 μM indexed P1/Universal P2 adapter mix from the prepared plate into each well of the plate containing dA-tailed samples from Subheading 3.2.5.
2. Make a reaction mix for 96 samples (prepare an additional 10% extra mix to ensure there is enough) containing the following components per sample:

Quick Ligase buffer	12.5 μL
Quick Ligase	0.5 μL

3. Pipet 13 μL of the reaction mix into the sample plate.
4. Cover the plate and incubate at room temperature (~20 °C) for 20 min, then heat inactivate at 65 °C for 5 min in a thermal cycler (*see* **Note 6**).
5. If performing the protocol without a library normalization step, proceed to Subheading 3.2.9.
6. If performing the protocol with a library normalization step, perform a SPRI bead clean-up as in Subheading 3.1.7, using 56 μL of beads (1.8:1) ratio, 200 μL for each ethanol wash, an elution volume of 30 μL of 10 mM Tris, and transferring 28 μL of eluate into a new 96-well microplate.

3.2.7 PCR Enrichment of Fragments (Only for Using the Protocol with Library Normalization)

1. Set up the PCR mix for 96 samples (prepare an additional 10% extra mix to ensure there is enough) using the following components per sample:

Water	13 μL
5× Phusion HF buffer	5 μL
10 mM dNTPs	0.5 μL
10 μM forward primer	0.25 μL
10 μM reverse primer	0.25 μL
50 mM $MgCl_2$	0.75 μL
Phusion DNA polymerase	0.25 μL

2. If intending to pool several 96-plexes in the same Illumina lane, use the universal forward primer and choose from the reverse indexing primers in Supplementary Table 3 for each 96-plex (*see* **Note 11**). These will be sequenced in the first indexing read if the Illumina machine is in dual-indexing mode or the sole indexing read if the machine is running in single-indexing mode.
3. Pipet 20 μL PCR mix into each well of a fresh 96-well plate. Save some of the leftover PCR mix for use in quantification.
4. Add 5 μL of purified, ligated DNA fragments from Subheading 3.2.6, **step 6**.
5. Cover plate and incubate in a thermal cycler at 98 °C for 30 s; 12 cycles of 94 °C for 10 s, 65 °C for 30 s, 72 °C for 30 s; 72 °C for 5 min; hold at 10 °C (*see* **Note 6**).

3.2.8 Quantification, Normalization and Pooling of Libraries

1. Quantify PCR amplicons using a Qubit fluorometer with the HS Kit or using a PicoGreen kit with a fluorescence plate reader. Dilute the standards in the PCR mix. This is to achieve more accurate quantification without having to do an additional bead clean-up step. Components of the PCR mix slightly alter the fluorescence readings, but this can be accounted for by mixing the standards with the PCR mix.
2. Normalize samples into a pool containing 20 ng of each library in a 1.5-mL microcentrifuge tube. Proceed to Subheading 3.2.10.

3.2.9 Pooling of Libraries Without Normalization (Skip this Step If Performing Subheadings 3.2.7 and 3.2.8)

1. Combine 8-14 μL of each library from Subheading 3.2.6, **step 5** into a single pool.
2. Proceed to Subheading 3.2.10.

3.2.10 Select Fragments from 200 to 500 bp Using SPRI Beads

1. Add SPRI beads at a ratio of 0.6:1. If library pool volume is >1 mL, then divide into two tubes of equal volume (*see* **Note 12**).
2. Mix and incubate at room temperature for 5 min.
3. Place tubes in a magnetic separation rack for 1.5-mL microcentrifuge tubes and wait 5 min for the sample to clear.
4. Transfer supernatant to a fresh tube. Note the supernatant volume. At this point, the 0.6:1 SPRI bead fraction can also be saved in case the size selection needs to be repeated (*see* **Note 13**).
5. Add SPRI beads at a ratio of 0.2:1 to the supernatant from **step 4** above.
6. Mix and incubate for 5 min.

7. Place tubes in a magnetic separation rack for 1.5-mL microcentrifuge tubes and wait 5 min for the sample to clear.
8. Remove supernatant and wash 2× with 700 μL 80% ethanol.
9. Allow the beads to dry completely at room temperature while the samples are still sitting on the magnetic stand (usually 5–20 min).
10. Remove tubes from the magnet and resuspend in 31 μL 10 mM Tris.
11. Mix and incubate for 5 min at room temperature.
12. Place tubes back on magnet and wait for approximately 5 min or until all of the beads are drawn to the magnet.
13. Transfer 30 μL to a fresh 1.5-mL microcentrifuge tube. Combine the two samples if the library pool was divided in **step 1**.
14. If the library normalization Subheadings in 3.2.7 and 3.2.8 were skipped, then proceed to Subheading 3.2.11. If the library normalization steps were included, then proceed to Subheading 3.2.12.

3.2.11 PCR Enrichment of Library Pool with Phusion Taq Polymerase

1. Prepare one PCR reaction for each 96-plex pool in a 50 μL total volume:

5× Phusion HF buffer	10 μL
10 mM dNTPs	1 μL
10 μM forward primer	1 μL
10 μM reverse primer	1 μL
Phusion Taq polymerase	0.5 μL
Library pool	3–15 μL
Water	add to total volume of 50 μL

2. If intending to pool several 96-plexes in the same Illumina lane, use the universal forward primer and choose from the reverse indexing primers in Table 4 for each 96-plex (*see* **Note 9** and **11**). These will be sequenced in the first indexing read if the Illumina machine is in dual-indexing mode or the sole indexing read if the machine is running in single-indexing mode.
3. Start with 3 μL of the library pool as input in the PCR reactions. This PCR enrichment step can be repeated with larger input volumes of the library pool if the final library yield after Subheading 3.2.12 is too low.

4. Perform PCR in a thermal cycler using the following conditions: 98 °C for 30 s; 12 cycles of 94 °C for 10 s, 65 °C for 30 s, 72 °C for 30 s; 72 °C for 5 min; hold at 10 °C (*see* **Note 6**).
5. Perform a bead clean-up of the PCR reaction using 98 μL of beads (1.8:1) ratio, 200 μL for each ethanol wash, an elution volume of 30 μL of 10 mM Tris, and transferring 28 μL of eluate into a 1.5-mL microcentrifuge tube (*see* **Note 6**).

3.2.12 Library Validation

1. Quantify the final library using the Qubit HS kit. Analyze the final library pool on a Bioanalyzer using a High Sensitivity chip or a DNA 1000 chip. Use the Bioanalyzer trace to calculate the size and nanomolarity. An example of a successful whole-genome 96-plex is shown in Fig. 3b.
2. Insert size should range from 200 to 500 bp. A final concentration of 10 nM is needed for the HiSeq 2000 or MiSeq platforms. For the HiSeq 3000, a final concentration of only 2.5 nM is needed.
3. Library concentration can be adjusted by dilution or by concentration using a standard SPRI cleanup with a 1.8:1 ratio and eluting in the volume required to achieve the desired concentration.
4. If the yield is too low, Subheadings 3.2.7 and 3.2.8, or 3.2.10 can be repeated.
5. If several 96-plexes are pooled into the same Illumina lane, then perform the validation and final concentration for each 96-plex library separately before mixing libraries in an equimolar ratio. To adapt any of the Subsections in Sections 3.1 and 3.2 for automation using a liquid-handling robot, *see* **Notes 14–17**.

4 Notes

1. The first five to eight sequenced bases make up the barcode, or index, of each read. Illumina recommends that all four bases (A, G, T, C) are equally represented in the first five positions of the read. Choosing barcodes where these positions are balanced ensures accurate cluster detection. To maintain the balance of bases, as few as 48 samples can be pooled. On all four mixed adapter plates, wells A01-H06 and A07-H12 can be combined column-wise to generate balanced libraries. We do not recommend pooling fewer than 48 samples. These recommendations are based on the well positions listed in Supplementary Table 1. We do not recommend pooling fewer than 48 samples, as the balance of bases will not be maintained. If pooling between 49 and 384 samples the barcodes can be used in order (Plate 1, then Plate 2, etc.).

2. These adapters only allow single-end sequencing. For paired-end sequencing, the oligo sequences must be modified. Paired-end sequencing can be enabled by changing the universal adapter oligos. However these oligos reduce the efficiency of ligation to the desired genomic fragments and use of these oligos is not recommended (PE-Universal_Adapter_1: AGATCGGAAGAGCGGGGACTTTAAGC and PE-Universal_Adapter_2: GATCGGTCTCGGCATTCCTGCTGAACC GCTCTTCCGATCTT).
3. If a small amount of ethanol remains, wait 1 min and then remove the residual liquid using a 10-μL multi-channel (or single-channel) pipet. This will ensure that the samples dry within 20 min. If small cracks begin to form in the bead pellet, this is a sign that the beads have become too dry and might be difficult to resuspend.
4. If beads do not become fully resuspended by pipetting then cover plate with a plate seal, vortex until all beads are in solution, and then spin liquid to the bottom of each well using a short centrifuge step.
5. All four plates are not combined by this step because the magnetic separation rack can only hold 2.0 mL tubes. If you have a magnetic separation rack that can hold larger volume tubes, then you may adjust this section. In case the 384 samples were prepared in smaller pools and not prepared all at the same time, we recommend measuring each of these smaller pools again after purification and combining them in an equal molar ratio in order to avoid uneven coverage across batches.
6. This is a safe stopping point. Samples can be frozen and thawed out before continuing if needed.
7. The next steps can be carried out in either a 1.5-mL microcentrifuge tube, 8-well strip tubes, or 96-well plate. With the 96-well magnet, it is possible to scale the preparation to include more than one 384-plex, so 8-well strip tubes or 96-well plates are preferred for this purpose.
8. We recommend doing a test PCR with two different cycles: 12 and 14. There is an excess of ligation product generated in Subheading 3.1.14 that can be used to test for the optimal cycle number. Choose the lowest cycle possible that generates bands covering the size range shown in Fig. 3.
9. The index sequences will be read as the first 8 bases of read 1 during sequencing on any Illumina machine. For optimal cluster detection, Illumina recommends that all four bases (A,G,T,C) should be equally represented in the first five positions of the read. If you order the indexed adapters according to the well positions listed in Supplementary Table 2, then you can pool samples so that they have the proper base balance in the first five positions if fewer than 96 samples are used. For example, if only

eight samples are pooled, then use only column 1. If 16 samples are pooled, then use columns 1 and 2. If 24 samples are pooled, use columns 1, 2 and 3. Proceed column-wise until you reach 96 samples.

10. The digestion time was optimized for 50–100 ng of *Arabidopsis thaliana* DNA. We recommend testing 100 ng with different digestion times (from 5 to 30 min) by running all of the sample on a gel to check size distribution if studying an organism other than *Arabidopsis thaliana*.
11. Pool at least four 96-plexes in a single lane using this method. If you want to pool more than four in a single lane, then pool them in the order listed in the "Order" column of Supplementary Table 3.
12. The reason for dividing the sample into two tubes is the capacity of the magnetic separation rack. If you have a magnetic separation rack that can accommodate larger tubes, then the library pool can be processed in one batch.
13. If the 0.6:1 SPRI fraction is to be saved, add 700 μL of 80% ethanol at this point. The second ethanol wash, drying step, resuspension and elution can then be performed in parallel with the 0.2:1 SPRI fraction in **step 8** of Subheading 3.2.10.
14. Hands-free automation of protocols with a liquid handling robot is attractive, but most robotic workstations are limited by deck space and do not have all necessary peripheral machines for full automation such as thermal cyclers. Consider a semi-automatic workflow, where large parts of the protocol are performed via several smaller robotic scripts (such as a script for reagent distribution, another script for SPRI cleanup), which is much simpler to organize and saves nearly as much time.
15. If the liquid handler being used has a limited number of pipetting channels, it may be difficult to finish certain steps within the recommended times, especially when slow pipetting is required due to viscosity of solutions such as the PEG-NaCl buffer used with SPRI beads. In SPRI bead cleanups, timings for all steps except drying steps are to be considered minimum incubation times, which can be safely extended as necessary; up to 30 min extra time in PEG-NaCl buffer, in ethanol, or in elution buffer, does not noticeably affect yield. It is critical to avoid extended periods where the SPRI beads are in empty wells, because yield can be reduced if they are allowed to dry excessively. For example, if washing SPRI beads with 80% ethanol in two 96-well plates, it is preferable to remove the supernatant from one plate and replace it with fresh 80% ethanol, before moving on to remove the 80% ethanol from the second plate.
16. Unwanted air pockets at the bottom of wells can lead to poor magnetic separation of SPRI beads and pipetting of incorrect

amounts. Consider faster dispense speeds for non-viscous solutions like ethanol. For viscous solutions where this is not possible, an additional bubble-destroying pipetting step is useful and highly effective; aspirate about half of the well volume from the bottom of the well, and dispense this volume (which will contain any air bubble that was in the well) on the liquid surface.

17. Optimize robotics for speed only to the extent that is possible without sacrificing quality; a major value of a properly programmed liquid handler is reliable precision and hands-free time, even if it is slower than a skilled bench scientist working manually.

Acknowledgements

We thank Paulo Teixeira for sharing his modifications of the Rohland and Reich (2012) protocol for preparing the Serapure SPRI beads, and George Wang for analyzing the index sequences for the 96 whole genome adapters and indexing primers to determine optimal base balance. We also acknowledge Norman Warthmann for designing the indexing primers.

References

1. Baird NA, Etter PD, Atwood TS et al (2008) Rapid SNP discovery and genetic mapping using sequenced RAD markers. PLoS One 3:e3376
2. Monson-Miller J, Sanchez-Mendez DC, Fass J et al (2012) Reference genome-independent assessment of mutation density using restriction enzyme-phased sequencing. BMC Genomics 13:72
3. Cao J, Schneeberger K, Ossowski S et al (2011) Whole-genome sequencing of multiple Arabidopsis thaliana populations. Nat Genet 43:956–963
4. Rowan BA, Patel V, Weigel D et al (2015) Rapid and inexpensive whole-genome genotyping-by-sequencing for crossover localization and fine-scale genetic mapping. G3 5:385–398
5. Rohland N, Reich D (2012) Cost-effective, high-throughput DNA sequencing libraries for multiplexed target capture. Genome Res 22:939–946

Chapter 17

Describing Sequence Variants Using HGVS Nomenclature

Johan T. den Dunnen

Abstract

DNA sequencing is usually performed to determine the sequence of a region of interest or even the entire genome of an individual. After sequencing, the sequence obtained is compared to a reference, all differences (the variants) are recorded, and the possible consequences of the changes identified, on both the RNA and protein level, are predicted. Finally, when available, a database containing previously reported variants is consulted to determine what other studies might have revealed about the variant or other variants in the same sequence (gene) and what the functional and phenotypic consequences were for the individuals carrying the variant.

To facilitate the reporting and databasing of variants a standard was developed, the HGVS recommendations for the description of sequence variants. HGVS nomenclature contains specific formats to describe the basic variant types; substitution, deletion, duplication, insertion, inversion, and conversion. The basics of how to apply the recommendations to describe sequence variants will be explained here. An extensive description of the current HGVS guidelines (version 15.11) is available online at http://www.HGVS.org/varnomen.

Key words Standards, Variant, Nomenclature, Database, Mutation, DNA, RNA, Protein

1 Introduction

The basics of the nomenclature system for the description of sequence variants proposed by the Human Genome Variation Society (HGVS) were published in a paper in 2000 [1]. The HGVS nomenclature was build upon some initial guidelines proposed since 1996 [2, 3], suggesting a comprehensive basic set of recommendations that has since developed into an internationally accepted standard. Applied in practice it was recognized the initial guideline had a few errors, contained some inconsistencies, and did not cover all variant types. An update was published recently, summarizing all changes since 2000, bringing the HGVS nomenclature to its current version, 15.11 [4].

The recommendations are currently under the auspices of three international organizations, the Human Genome Variation Society (HGVS), the Human Variome Project (HVP), and the Human

Stefan J. White and Stuart Cantsilieris (eds.), *Genotyping: Methods and Protocols*, Methods in Molecular Biology, vol. 1492, DOI 10.1007/978-1-4939-6442-0_17, © Springer Science+Business Media New York 2017

Genome Organisation (HUGO), and commissioned through a Sequence Variant Description Working Group (SVD-WG). Requests for changes go through the SVD-WG following a standard procedure including a final community consultation step. To make sure you have the opportunity to give your opinion, the suggestions is to register for the SVD-WG mailing list: mail to Tim Smith (tim@variome.org).

The description of sequence variants as presented in this Chapter are based on the current HGVS recommendations, version 15.11 [4]. The focus will be on the description of simple variants, for the description of complex variants, which in general are rare, we refer to the HGVS nomenclature web site (http://www.HGVS.org/varnomen). The nomenclature pages list an email address for questions (VarNomen@HGVS.org). To promote the recommendations a Facebook page (http://www.facebook.com/HGVSmutnomen) has been initiated where, on a regular basis, topics of interest are discussed.

2 Materials

Besides the HGVS recommendations, as explained extensively on the HGVS website (http://www.HGVS.org/varnomen), the Mutalyzer suite [5] (available from http://www.Mutalyzer.nl) will be used to check and/or generate HGVS variant descriptions. Other free software tools that can be used include a Python package for parsing, validating, mapping, and formatting sequence variants using HGVS nomenclature [6] and the Variant Validator (http://variantvalidator.org). Following HGVS recommendations we will not use the confusing terms "polymorphism" and "mutation" but neutral terms like "variant", "alteration", and "change" only.

3 Methods

To prevent confusion HGVS nomenclature includes specific definitions of the different basic variant types recognized; substitution, deletion, duplication, insertion, inversion, and conversion. In addition, variant types have been prioritized. When a variant can be described using several classes, the priorities are; (1) substitution, (2) deletion, (3) inversion, (4) duplication, (5) conversion, (6) insertion. Consequently, when a variant can be described as an inversion or substitution it should be described as a substitution. Similarly, when a variant can be described as a duplication or an insertion, duplication is preferred.

3.1 Reference Sequence

Step 1 to describe a variant is the selection of a reference sequence, a master/template considered to contain the normal ("wild type") sequence (*see* **Note 1**). Only reference sequences publicly available from the major DNA databases (EBI, NCBI, DDBJ) are accepted. HGVS recognizes several reference sequence types, indicated using a "prefix" preceding the variant description. Approved references are a genomic (g. prefix), mitochondrial (m.), coding DNA (c.), noncoding DNA (n.), RNA (r.), and protein (p.) reference sequence. The reference used should represent the sample analyzed; when genomic DNA is used a genomic reference sequence is referred, when RNA (cDNA) was analyzed a RNA reference sequence. The preferred reference is a (chromosomal) genomics reference sequence derived from a recent genome build; for human GRCh37/hg19 (e.g. NC_000011.9 for chromosome 11) or GRCh38/hg38 (NC_000011.10 for chromosome 11). For diagnostic applications HGVS strongly recommends the Locus Reference Genomic sequence (LRG) [7]. When no LRG is available, one should be requested as soon as possible. In the mean time, a RefSeq sequence [8], with its version (RefSeqGene or transcript) is recommended. The reference sequence used must contain the variant residue described.

HGVS nomenclature uses IUPAC-IUB symbols to describe nucleotides (DNA and RNA) and amino acids [9]. In protein variant descriptions the use of the three-letter amino acid code is recommended above the one-letter code, and "Ter" or "*" are used to indicate a translation termination (stop) codon (TAA, TAG, or TGA).

3.2 Numbering

Residue numbering (nucleotides and amino acids) in a reference sequence is straightforward, starting with 1 from the first to the last nucleotide, e.g. g.1, g.2, g.3, …, etc. and numbers do not include "+", "–", "*", or other preceding signs. The only exception is a coding DNA reference sequence (with associated RNA reference) where numbering starts with "c.1" at the A of the ATG-translation initiation (start) codon and ends with the last nucleotide of the translation termination (stop) codon (i.e. TAA, TAG, or TGA). Nucleotides upstream (5′) of the ATG are marked with a "–" and numbered c.–1, c.–2, c.–3, etc. (going upstream), nucleotides downstream (3′) of the stop codon are marked with a "*" and numbered c.*1, c.*2, c.*3, etc. (downstream). Nucleotides at the 5′ end of an intron are numbered relative to the last nucleotide of the directly upstream exon, followed by a "+"and their position in to the intron (c.87+1, c.87+2, c.87+3, …), nucleotides at the 3′ end of an intron are numbered relative to the first nucleotide of the directly downstream exon, followed by a "+" and their position out of the intron (…, c.88–3, c.88–2, c.88–1).

A range of nucleotides is indicated using a "_", flanked by the first and last nucleotide involved; g.80_88 meaning nucleotides g.80 to g.88. For variants in stretches of repeated DNA sequences HGVS assigns by definition the most 3′ position possible as being changed. This so called 3′ rule means the change of TTT to TT is described as g.3del (not as g.1del or g.2del). At the protein level, deletion/duplications of amino acids are similarly normalized to the most C-terminal position.

3.3 Mutalyzer

The hypothetical variant descriptions given below will all be based on the chromosomal genomic reference sequence NC_000022.10, i.e. human chromosome 22 genome build GRCh37/hg19, and the LRG_989t1 coding DNA reference sequence of the *LZTR1* gene (based on NG_034193.1/NM_006767.3). Variant descriptions were all checked with the Mutalyzer tool (http://www.Mutalyzer.nl/). Table 1 contains a reference and a variant sequence which can be used to generate, using Mutalyzer's Variant Description Extractor (http://www.Mutalyzer.nl/description-extractor) [10], all example variant descriptions given below. Simply copy/paste the two sequences in the respective Mutalyzer fields and click the "Extract variant description" button. Numbering for "c." is correct, for "g." add 21,336,660.

3.4 DNA Variant Descriptions

3.4.1 Substitutions

A substitution is defined as a sequence change where, compared to a reference sequence, one nucleotide is replaced by one other nucleotide. Substitutions are indicated using a ">". Substitutions are described using the format NC_000022.10:g.21336689A>C; a substitution of the A nucleotide at position g.21336689 for a C. Based on a coding DNA reference sequence LRG_989t1:c.29A>C.

3.4.2 Deletions

A deletion is defined as a sequence change where, compared to a reference sequence, one or more nucleotides are not present (deleted). Deletions are indicated using "del". One-nucleotide deletions are described using the format g.21336738del/c.78del (alternatively g.21336738delG/c.78delG); a deletion of the G nucleotide at position g.21336738 (c.78 based on a coding DNA reference sequence). Several-nucleotide deletions are described using the format g.21336785_21336805del/c.125_145del); a deletion of nucleotides g.21336785/c.125 to nucleotide g.21336805/c.145.

3.4.3 Duplications

A duplication is defined as a sequence change where, compared to a reference sequence, a copy of one or more nucleotides are inserted directly 3′ of the original copy of that sequence (tandem copy). Duplications are indicated using "dup". One-nucleotide duplications are described using the format g.21336731dup/c.71dup (alternatively g.21336731dupT/c.71dupT); a duplication of the T nucleotide at position g.21336731/c.71. Several-nucleotide duplications are described using the format g.21336821_21336829dup/c.161_169dup; a duplication of nucleotides g.21336821/c.161 to nucleotide g.21336829/c.169.

3.4.4 Insertions

An insertion is defined as a sequence change where, compared to the reference sequence, one or more nucleotides are inserted and where the insertion is not a copy of a sequence immediately 5′.

Table 1
Sequences to generate the coding DNA variant descriptions mentioned as example in the main text using the Variant Description Extractor form Mutalyzer (http://www.mutalyzer.nl/description-extractor) [10]

Sequence	Copy/paste to Mutalyzer window
Reference sequence	ATGGCTGGACCGGGCAGCACGGGGGGGCAGATCGGGGCTGCGGCCCTGGCAGGCGGCGCGGTCCAAGGTAGCCCCGA GCGTGGACTTCGACCATAGCTGCTCGGACAGTGTCGAGTACCTGACGCTCAACTTCGGGCCCTTCGAAACAGTGCATCGC TGGCGGCGCCTCCCGCCCTGC
Sample sequence	ATGGCTGGACCGGGCAGGTGGGGGGGCCGATCGGGGCTGCGGCGGTCCCTGGCAGGCGGCGCGGTCCAAGGTTAGC CCCAGCGTGGACTTCGACCATAGCAACACTCGGACAGTGTCGAGTACCTGAAAACAGTGCATCGCTGGCGGCGCCGGCGG CGCCTCCCGCCCTGC

The sequences were copied from genomic reference sequence LRG_989t1, nucleotides c.1 to c.120. based NG_034193.1 and NM_006767.3

Insertions are indicated using "ins" and described using the format g.21336704_21336705insGGTC/c.44_45insGGTC; an insertion of nucleotides GGTC between nucleotides g.21336704/c.44 and g.21336705/c.45.

3.4.5 Inversion

An inversion is defined as a sequence change where, compared to a reference sequence, more than one nucleotide replacing the original sequence are the reverse complement of the original sequence. Inversions are indicated using "inv" and described using the format g.21336678_21336680inv/c.19_21inv; an inversion of nucleotides CAC from position g.21336678/c.18 to 21336680/c.20 to GTG.

3.4.6 Conversion

A conversion is defined as a sequence change where, compared to a reference sequence, a range of nucleotides are replaced by a sequence from elsewhere in the genome. Conversions are indicated using "con" and described using the format NC_000022.10:g.42522624_42522669con42536337_42536382 (hg19); a conversion of chromosome 22 nucleotides g.42522624 to g.42522669, replacing them with nucleotides g.42536337 to g.42536382 (*CYP2D7* gene conversion in exon 9 of *CYP2D6*, not present in the Mutalyzer VDE example).

3.4.7 Complex

Complex changes are defined as a sequence change where, compared to a reference sequence, a range of changes occur that can not be described as one of the basic variant types (substitution, deletion, duplication, insertion, conversion, inversion). Complex changes go from simple compound variants like deletion-inversions ("indels") and repeat sequence variability, to balanced and unbalanced translocations, marker chromosomes, and chromothripsis events. In general such variants are rare and we refer to the HGVS nomenclature pages for recommendations how to describe them (http://www.HGVS.org/varnomen). Indels are described as a deletion followed by an insertion using the format g. g.21336760_21336761delinsAACA/c.100_101delinsAACA; a deletion of nucleotides TG from position g.21336760/c.100 to g.21336761/c.101 being replaced by AACA.

3.5 Alleles

To describe several variants on the same allele (chromosome) or on the two different alleles, variants are grouped between square brackets ("[]"). c.[29A>C;78del] describes two variants known to be on one molecule (in cis). c.[44_45insGGTC];[161_169dup] describes two variants known to be on two different molecules (in trans). When the phase is not known, the description has the format c.[29A>C(;)19_21inv].

When at the RNA level two different transcripts are detected that derive from one allele this is described using the format r.[76a>c, 73_88del]; the nucleotide change c.76A>C yields two RNA molecules, one carrying variant 76a>c and one containing a deletion of nucleotides 73 to 88, caused by a shift of the splice donor site to within the exon.

3.6 Repeated Sequences

A repeated sequence is defined as a sequence where, compared to a reference sequence, a repetition of one or more nucleotides is variable. Repeated sequences can be any size, varying from small (mono-, di-, tri-, etc., nucleotide) to kilobase-sized repeats. g.123_125[36] (or g.123GGC[36]) describes a repeated sequence with the first repeat-unit located at position g.123 to g.125, present in 36 copies in the sample analyzed. As shown, g.123GGC[36], an alternative description based on the sequence of the repeat unit is allowed, however for larger repeat units this format quickly becomes impractical. In addition, when the description g.123GGC[36] is given it means the repeat was actually sequenced. g.123_125[36] does not specify the repeat sequence and can be used when the size of the repeat was determined using e.g. gel electrophoresis.

3.7 RNA Variant Descriptions

Variant descriptions on the RNA level largely follow those at the DNA level. RNA-level descriptions are only given when RNA (cDNA) has been sequenced. RNA variants are described using a coding or noncoding RNA reference sequence with nucleotide numbering copied from the "c." or "n." reference sequence. Nucleotides at the RNA level are given in lower case and the "T" is replaced for a "u" (uracil).

When RNA has not been analyzed the change at RNA can be best described as "r.(?)"; RNA was not analyzed but the change is expected to be identical to that a the DNA level. Exceptions are variants that most probably alter splicing which can be described as "r.spl" (variants at +1, +2, −2, −1 intron positions). Similarly when an effect on splicing is possible but not sure this can be indicated using r.(spl?), e.g. variants at the first or last nucleotide of an exon, at intron positions +3, +4, +5, and others. When the promoter/transcription start of a gene is deleted the predicted absence of a transcript can be indicated using "r.0?" (note that a new promoter/transcription start site could be activated).

3.8 Protein Variant Descriptions

Amino acids are preferably described using the three-letter amino acid code. The translation termination codon is described as "Ter" or "*". Descriptions should clearly show whether experimental proof was available or whether the description given is simply a prediction based on the change found at the DNA level. Predicted consequences should be listed in parentheses, e.g. p.(Arg23Ser).

Variant descriptions at the protein level start with the amino acid affected, followed by its codon number and then a description of the actual change. Substitutions have the format p.(Arg23Ser) and do not, as on DNA and RNA level, use the ">" (i.e. not p.Arg23>Ser). Deletions are described as p.(Arg23del) or p.(Trp45_His53del), duplications as p.(Arg23dup) or p.(Trp45_

His53dup), insertions as p.(Lys33_Leu34insHis), conversions as p.(Asn34_Gln134conAsp302_His402). Inversions are not used on the protein level.

Variants that are predicted to shift the translational reading frame are described using either a short or a long form; p.(Arg97fs) or p.(Arg97Profs*23). In"fs*23", "23" indicates at which codon number the new reading frame ends with a stop codon. The description p.(Ala127Profs*1) is not possible, correct is either p.(Ala127Profs*2) or p.(Ala127*).

3.9 Uncertainties

HGVS allows the description of uncertainties using parentheses (as in predicted protein consequences) and a "?". The format is used most frequently to describe deletion breakpoints detected using techniques like MLPA, PCR, or arrays. In the description the range of the uncertainty is listed in parentheses, e.g. (5′ border_3′ border), describing the change as precise as possible. g.(1234_3456)_(5678_7890)del describes a deletion with neither breakpoint sequenced, but defined to be between g.1234 and g.3456 on one side and between g.5678 and g.7890 on the other side. When for a repeated sequence the size of the repeat expansion is uncertain, this is be described as g.-128_-126[(600_800)]; the tri-nucleotide repeat has between 600 and 800 copies.

4 Notes

1. Looking at publications containing variant descriptions, the most commonly made mistakes are:

   (1) The reference sequence used is not mentioned or its version number is lacking (see Subheading 3.1, e.g. NM_006767.3),

   (2) The 3′ rule is not applied properly, i.e. in repeated sequences variant descriptions are not shifted as far 3′ as possible (see Subheading 3.2),

   (3) Duplications are described as insertions, neglecting the rule of variant prioritization (see Subheading 3).

   (4) Intronic variants are described in relation to a reference sequence not containing the variant nucleotide (see Subheading 3.1, e.g. NM_006767.3:c.200+1G>A)

   (5) Predicted protein descriptions are not given in parenthesis (see Subheading 3.8, correct is p.(Arg45Ser))

   (6) The translation termination codon (stop) is described using an "X" (see Subheading 3.8, correct is "Ter" or "*")

   (7) Variants in introns are described using 'IVS1+1G>A', neglecting the rule to use nucleotide numbers only

(8) Insertions are described as c.123ins3 neglecting the rule to describe between which nucleotides the insertion is and not specifying the sequence inserted (see Subheading 3.4, correct is c.123_124insAGG)

(9) Deletion/duplication ranges are erroneously shortened to c.123-93_-69del (see Subheading 3.2, correct is c.123-93_123-69del)

(10) It is not clearly described whether variants are found on the same or on different alleles or, in recessive disease cases, which variants were found in which combination (see Subheading 3.5).

Acknowledgements

This chapter was written on behalf of the HGVS/HVP/HUGO Sequence Variant Description Working Group.

References

1. Den Dunnen JT, Antonarakis SE (2000) Mutation nomenclature extensions and suggestions to describe complex mutations: a discussion. Hum Mutat 15:7–12
2. Ad Hoc Committee on Mutation Nomenclature (1996) Update on nomenclature for human gene mutations. Hum Mutat 8:197–202
3. Antonarakis SE (1998) Recommendations for a nomenclature system for human gene mutations. Hum Mutat 11:1–3
4. Den Dunnen JT, Dalgelish R, Maglott DR et al (2016) HGVS recommendations for the description of sequence variants: 2016 update. Hum Mutat 37:564–569
5. Wildeman M, van Ophuizen E, den Dunnen JT et al (2008) Improving sequence variant descriptions in mutation databases and literature using the Mutalyzer sequence variation nomenclature checker. Hum Mutat 29:6–13
6. Hart RK, Rico R, Hare E et al (2014) A Python package for parsing, validating, mapping and formatting sequence variants using HGVS nomenclature. Bioinformatics 31:268–270
7. Dalgleish R, Flicek P, Cunningham F et al (2010) Locus Reference Genomic sequences: an improved basis for describing human DNA variants. Genome Med 2:24.1–24.7
8. O'Leary NA, Wright MW, Brister JR et al (2016) Reference sequence (RefSeq) database at NCBI: current status, taxonomic expansion, and functional annotation. Nucleic Acids Res 44:D733–D745
9. IUPAC-IUB Joint Commission on Biochemical Nomenclature (1984) Nomenclature and symbolism for amino acids and peptides. Recommendations 1983. Eur J Biochem 138:9–37
10. Vis JK, Vermaat M, Taschner PE, et al (2015) An efficient algorithm for the extraction of HGVS variant descriptions from sequences. Bioinformatics 31:3751–3757

INDEX

Stefan J. White and Stuart Cantsilieris (eds.), *Genotyping: Methods and Protocols*, Methods in Molecular Biology, vol. 1492, DOI 10.1007/978-1-4939-6442-0, © Springer Science+Business Media New York 2017

MIX
Papier aus verantwortungsvollen Quellen
Paper from responsible sources
FSC® C105338

If you have any concerns about our products,
you can contact us on
ProductSafety@springernature.com

In case Publisher is established outside the EU,
the EU authorized representative is:
Springer Nature Customer Service Center GmbH
Europaplatz 3, 69115 Heidelberg, Germany

Printed by Libri Plureos GmbH
in Hamburg, Germany